Life, Liberty
and The Pursuit of
Health Care

Life, Liberty

and The Pursuit of

Health Care

by Edwin Tutt Long, M.D.

with Ann Buzenberg

Published by Lulu.com

Cover design by Annie Pursley

Book design by Ann Buzenberg

ISBN 978-1-4357-1191-4

CONTENTS

Dedicated to

William Bassett, Director of Planning and Assessment,

and

Mike Tansey, Ph.D., Professor of Economics,

Rockhurst University

and to

my beloved wife Mary

PREFACE

Greetings!

That was the way my draft board informed me, years ago, that my country needed me. Today it's the way I'm telling you your country needs you. This time, the United States of America needs you to help mobilize an outcry and call for universal health care coverage.

But wait! you say. The call for universal health coverage is not new. The pursuit of health care in our country has been debated and discussed for decades. So why am I calling on you now?

The demand for health care change is building like a perfect storm. This is because multiple parts of our system are reaching a crisis.

In 2006, 47 million Americans had no health care because they were priced out of the insurance system. And for the insured, the actual quality of our health care is in question: medical care prices are skyrocketing while good outcome rates are disappointing.

These numbers are forming a tsunami, pushing businesses, insurance companies, medical providers and the government to change the system, now. This tsunami is good. Its relentless pressure for change means that we are poised for universal health care to be enacted at last.

But change for the better is not a sure thing. That is why you and other informed American citizens are needed—to insist on coverage that benefits everyone, that allows excellence in health care and at a price that makes sense. While these goals may sound overblown, let me assure you they are very possible, and with surprisingly modest changes to the current systems now in place.

Furthermore, health care is your right as an American citizen, just like you are entitled, by being an American citizen, to a public education, kindergarten through grade 12, and to vote for your government representation.

As a retired physician with no agenda and no financial support from any organization, I have attended many high-level conferences around the nation where all points of view about our health care future were aired and debated. As I listened, one thing became clear to me. As a leading nation, it is unacceptable that the United States lets almost

50 percent of its population fend for itself in terms of health care and staying healthy. Other nations can guarantee health care for all their citizens. It is my conviction that we can do it too.

I will show you in detail how this can be done, so stay with me.

First in *The Health Care We Have*, we'll visit and understand our present health care system. We'll look at the major structural elements which actually create the problems unique to our current U.S. health care system. Understanding how these flaws were formed will help us recognize and reject proposed solutions which may be similarly flawed.

Next, in *The Health Care We Need*, we'll discuss possible solutions plus new directions that can manage, avoid or overcome unwanted results while giving us the full health benefits and coverage we want and need.

Changing the health care system is a prominent issue for politicians and others who offer their versions of how it should change. Their proposals seem to sort themselves into two groups. The first group calls for revising our present incremental system to reduce the number of uninsured but not necessarily covering everyone. The second group proposes a new system designed to offer health care to every American. This book falls in the second group.

Edwin Tutt Long, M.D.
Kansas City, Missouri

The great contribution we can make is to prepare the
oncoming generations to think
that they can and will think for themselves.

Charles H. Mayo, co-founder, Mayo Clinics

ACKNOWLEDGEMENTS

No man is an island, least of all a retired surgeon who has taken on the task of writing a book.

I am grateful to have been surrounded by a sea of astute, informed and wise colleagues and friends who unstintingly shared their precious time and even more valuable thoughts and insights with me along the way.

Brian Klepper, national executive director, and Mike Wood, regional co-chair of the Kansas City Chapter, Center for Practical Health Reform, and Len Nichols, New America Foundation, all provided invaluable guidance and assistance.

Professor Richard Frank of Harvard University, started me on the path that led to this book with his lecture about adverse selection.

I also owe a debt for his motivating advice to Dr. James Olsen, now-deceased president of the University of Missouri. When we met years ago, Dr. Olson listened, looked me squarely in the eyes, and said, "There are two questions. First: Is it a good idea? And second: Can you do it?"

For their comments, advice and wisdom, I wish to offer my deepest thanks to Peter Long, Laura Long, Mary Long, John Hood, Terry Roselle, Ronald Aractingi, Esq., Pat Friesen, Tim Wurst, William Bassett, and Mike Tansey.

Finally, to Ann Buzenberg, my editor and publishing advisor, I owe special gratitude for the hours she spent weaving my years of notes and writings into a book.

All errors and omissions are mine alone.

If wealth is measured by the value of the company we keep, I am indeed very rich.

Ed Long

If you don't know where you are going,

any road will get you there.

The Red Queen, *Alice in Wonderland*

PART 1

The Health Care We Have

Difficulties are meant to rouse, not discourage.

The human spirit is to grow strong by conflict.

William Ellery Channing (1780 - 1842)

CHAPTER 1

It's The System

America's present incremental health care system gives us our problems. In fact, the system is ideally set up to reap the poor results we are getting.

Incremental health care means that some people get their health care in one way and others get it in other ways. Some health care in the United States is public, such as through Veterans Administration hospitals and doctors. Other health care is available through private physicians and hospitals. Forty-seven million people under age 62 are uninsured.

I challenge you to find one American who is unaware that our health care system is both magnificent in its best accomplishments— like delicate microscopic surgery to restore nerve connections in severed limbs—and terribly flawed in its worst failures, such as high rates of sick patients who are made even sicker by hospital-based infections, and so many uninsured.

The U.S. in 2006 spent more than $2 trillion or $7,000 total on each person for health care, according to the report in the *New York Times*, quoting the Center for Medicine and Medical Research. Health care spending is 4.3 times the amount spent on national defense. The predicted health care expenditure in 2007 is $7,498 for each person in the U.S.

Health care expenditure is projected to reach $4 trillion in 2015.

A federal study released in September 2007 documented the variations in personal health care spending among the states. Utah spent less than $4,043 per person compared to Massachusetts that spent almost $7,075 per person in 2004. Or to put it another way, Utah residents spent 57 percent as much on health care as Massachusetts residents. Much of this was due to different styles of practice by physicians. Some of it was due to differences in employer-based health insurance benefit packages.

U.S. life expectancy at birth is now 77.9 years, according to the Centers for Disease Control. In contrast, France spent $3,337 per person

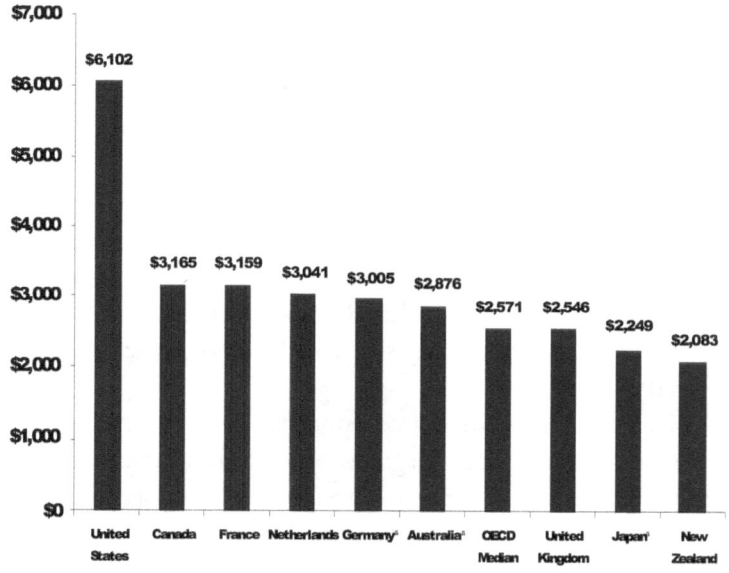

Health Care Spending per Capita in 2004
Adjusted for Differences in Cost of Living

J. Oylus and G. F. Anderson, *Multinational Comparisons of Health Systems Data, 2006*   (New York: The Commonwealth Fund, Apr. 2007).

on health care, for an expected lifespan of 80.3 years. Only Mexico, with an expenditure of $666 per person, had a lower life expectancy of 75.2 years.

Commenting about the wide variation in prices for the same health care services across the U.S., Professor Ewe Reinhardt of Princeton University asked, "How can the best health care in the world cost twice as much as the best health care in the world?"

Other countries are getting better benefits for all their citizens at a lower expense, and the U.S. is not even covering everyone. And, our problems are getting worse. Aren't we entitled to ask, "If we are spending 2.2 trillion dollars on health care, shouldn't we get good health outcomes equal to the European countries that spend considerably less than we do?"

A majority of Americans say the federal government should guarantee health insurance to every American, especially children, and are willing

Majority of Americans Believe Paying for Health Insurance Should Be a Shared Responsibility

Who do you think should pay for health insurance?

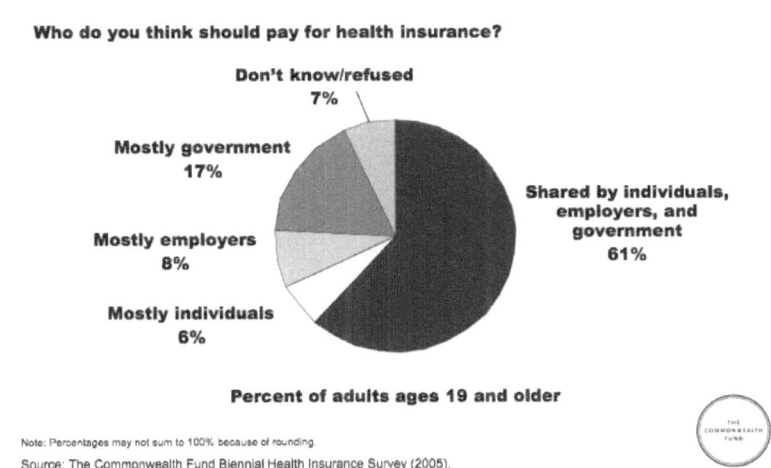

Percent of adults ages 19 and older

Note: Percentages may not sum to 100% because of rounding.
Source: The Commonwealth Fund Biennial Health Insurance Survey (2005).

to pay higher taxes to do it, according to a recent *New York Times*/CBS News poll.

As reported in the March 2, 2007, *New York Times*, "While the war in Iraq remains the overarching issue in the early stages of the 2008 campaign, access to affordable health care is at the top of the public's domestic agenda, ranked far more important than immigration, cutting taxes or promoting traditional values."

In this poll, Americans were willing make tradeoffs to guarantee health insurance for all, including paying as much as $500 more in taxes a year and forgoing future tax cuts.

The Heart of the Matter

So America's willingness to sacrifice for a change in order to get universal coverage is not the stumbling block that some think. However, any changes that we Americans embrace must address and solve our systemic health care problems.

To do this, you and I must understand the problems and then choose among the solutions—some will work and others will not. Understanding comes from knowing how the system works and who the players are within the system. Armed with the right understanding, our mandate for change can be effective.

CHAPTER 2

The Players

Who are the players in the health care system? We'll identify the players—also called stakeholders—in a minute. Just know, for now, that you and I are among them.

"All stakeholders are part of the problem," said Bruce Bradley, General Motors' director of health care strategy and public policy. "We need to take ownership and work together on the solutions." So let's get started.

The stakeholders, or players, have independent interests that, taken together, create the problems. This results in a health care system that allows, supports, and even rewards these key roles even while they cause problems.

The players are:

1 **American Public** whose good health is the goal.
2 **Health Care Providers**: Doctors, hospitals, and health care professionals. (Nurses generally work for one of these providers and so are not a "player" in the sense of causing problems by being paid directly.)
3 **Insurance Companies** which manage risk
4 **Governments**: State and local governments pay 13 percent and the federal government pays for 32.4 percent of U.S. health care.
5 **New Makers**: My term for innovators and up-daters in medicine, pharmaceuticals, technology and policy
6 **Employers**: 54 percent of U.S. health care in the U.S. is through employers voluntarily offering health insurance to employees.

Let's look at these stakeholders in our health care system. What are their self interests and their contributions to U.S. health care?

PLAYER 1: American Public

Theoretically, ideally, and ethically, the American people and their health are the focus and reason for our health care system. The system should exist for the peoples' health.

Yet people's health is not served very well by the system. That service is beset by many glaring problems—some you and I can control, and some created by other players in the system. Let's look at some of the biggest problems.

First and most obvious is the burgeoning number of Baby Boomers who will routinely qualify for social security and Medicare. Combine that with the increase in longevity—which is fortunate for people but unfortunate for the health care system—and you can see that a large percentage of our population is getting older, generating increasing medical expenses over a longer time span.

Time also has changed our lifestyles. A century ago, a large percent of Americans worked as farmers living a simple, physically vigorous lifestyle. Today, fewer than 3 percent of Americans are farmers. Most people now live complex urban lives with high stress, little physical exercise, and poor diets; these predispose us to develop such common diseases as obesity, hypertension, diabetes and coronary artery disease. All things considered, today's younger generation may be the first to live shorter lives than their parents, unless things change.

Added to the shifts in population age and lifestyle, now fully 40 percent of all premature deaths in United States are triggered by what experts consider preventable lifestyle behaviors. Nearly 30 million Americans suffer chronic diseases and one million die annually from entirely preventable causes. For example, the most important contributing factor for heart disease, cancer, stroke, respiratory disease, diabetes and accidents— 6 of the 10 leading causes of death in United States—is lifestyle choice, including poor dietary choices, obesity, smoking, lack of adequate exercise, alcohol and drugs.

Each year, more than 700,000 Americans go to the emergency room with harmful reactions to medicines. It is possible that even more adverse reactions to pharmaceutical drugs occur but may be attributed to something else. Accidental overdoses and allergic reactions to prescription drugs were the most frequent cause of a serious illness. The most common offenders were overdoses of insulin, warfarin and amoxicillin.

These are drugs that are quite effective but need to be managed very carefully.

Another problem is that we don't generally understand how our health care system works and how much it costs. Contributing to this lack of understanding is the employer who, in our present system, is virtually forced to offer health insurance as a benefit to employees. The employer is well aware of health care costs because his insurance rates have skyrocketed. Many employers are doing a better job of sharing the pain—that is, informing employees about costs and how they are affected. But do employees understand that this means higher percentages of health care costs are being transferred to them?

Even more troubling, people fail to follow instructions that accompany their medicines and treatment plans. First, they may not be able to afford to follow the instructions. Second, confusion can rise from the way instructions are written by the drug companies and handed out by the medical profession. The National Center for Educational Statistics found that five out of six adults complained that health care instructions are written in a way that is foreign to how people talk. A majority of U.S. adults have some difficulty using health-related materials with accuracy and consistency. Of course, literacy varies by socio-economic, ethnic and age groupings. But the bottom line is, low health compliance leads to higher costs, poor health outcomes and more emergency room visits.

Just what does it take to make us healthier? And, are these factors under our control? Let's look at Colorado, which developed a health care report card for its citizens. Colorado children received a C and adults received a B+. The scores were based on these and other health indicators:

- Smoking
- Obesity
- Screening for blood cholesterol
- Vaccination rate
- Physical activity
- Health insurance coverage
- Unintended teen pregnancy
- Adolescent depression

Not surprisingly, many of these can be controlled by our life choices.

Considering all the problems we've just explored, how satisfied are people with health care? A recent poll by *USA Today*, ABC News, and the Kaiser Family Foundation showed that:

- 51 percent of all insured people were dissatisfied with the quality of the U.S. health system.
- The uninsured registered 75 percent dissatisfaction.
- Fully 80 percent of all respondents said they were dissatisfied with the more than $2 trillion spent on health care each year.
- 25 percent of the wealthiest Americans and 50 percent of workers in low and middle range compensation jobs report problems with medical bills in any 12 month period.
- Medical problems for just one family member can affect the economic stability of the whole family.

Many Americans are demonstrating their dissatisfaction by voting with their wallets. In 2005, more than 500,000 people sought medical treatment outside the U.S., according to the National Coalition on Health Care. Many Americans choose surgery and other treatments in Thailand, India and Singapore. In Thailand, the Bumrungrad hospital in Bangkok cared for 55,000 Americans in 2005 and 64,000 in 2006. About 100 facilities in six Asian countries are accredited by The Joint Commission on Accreditation of Health Care Organizations.

Because of high costs in the U.S., even employers are choosing to send their employees outside the U.S. for medical care. Blue Shield of California and Health Net of California allow their members to seek medical care in Mexico.

The American Association of Retired Persons (AARP) has examined and written about going overseas for medical care. They list these advantages:

- Lower prices for medical and dental care.
- State-of-the-art hospitals.
- Western trained physicians and dentists who speak English.
- Faster access to doctors than in the United States.
- Vacation opportunities and post-care rest at a relatively low cost.

AARP also notes some disadvantages:

- Pre-trip anxieties such as, *Will they speak English? Am I well enough to travel? Will I be safe?*

- Lack of follow-up care by the foreign surgeon, should complications occur after the patient returns home
- Limited or no legal recourse in case of medical or hospital negligence
- Feelings of loneliness or strangeness, if making the trip alone
- Risk of poor treatment by under-qualified practitioners or in substandard facilities.

PLAYER 2: Providers

Providers are a complicated lot. They include physicians, nurses, hospitals, affiliated medical personnel and others. America needs providers. They are the trained, talented source of our medical and surgical support. They are the heart and power of the health care engine. Through the proper execution of their duties, they deserve our respect and financial reward. The model for health care is and always will be a doctor and a nurse, with help, caring for a patient.

Physicians

Physicians are providers with a unique set of problems. Not the least of these is that the doctor's professional knowledge base is constantly changing and updating. A busy physician in private practice has his or her hands full taking care of patients. Yet he or she also must attempt to stay informed of the latest clinical trial results, epidemiological findings, and best practice guidelines. That's a huge task and surely can result in a distracted physician.

Furthermore, a doctor is not rewarded on the basis of outcomes, but only for units of service, as we've seen. Thus, it comes as little surprise that the Institute of Medicine, in 2003, found somewhere between 30 and 50 percent of patient health care was inappropriate and wasted. We know that approximately 20 percent of laboratory assessments are repeats because the original results were lost. It is estimated that about 40,000 deaths occur annually in America due to medical errors.

Perhaps contributing to these errors is an apparent lack of communication, doctor to patient. As Larry Arnhart, professor of Political Science at Northern Illinois University and proponent of evolution, observes, human beings are different from other animals. Humans possess language, which gives them the ability to reflect upon and rank their desires. According to *The Archives of Internal Medicine*, today's physicians fail to communicate with their patients about why a drug is prescribed, when it should be stopped and what adverse effects might occur.

Another problem for physicians is that our present payment system encourages procedures over prevention and outcomes. As we've discussed earlier, the social arrangement today is to pay providers for units of work, called fee-for-service. The fee-for-service model results in billing for a number of services at quite expensive fees. If we charted a cost-benefit curve, it would reveal that we are far out on the flat of the curve. In comparing our health care benefits and costs with the same costs and benefits shown by other countries' universal health care, the price we pay is too high for the health care benefits we receive.

Unfortunately, people also expect their providers to deliver health care whenever an urgent need arises. They are surprised to find that their physician sends them to the emergency room. Providers are in the health maintenance business and frequently have no openings for urgent problems.

Professionalism

Providers earn their living by providing care to people, which is a business venture. At the same time, providers are constrained by professional Codes of Ethics to put patients' interests first. But in our capitalistic society, vexing problems arise from a medical provider's effort to satisfy both business requirements and medical ethics. Herein lies a core challenge for providers in today's health care system.

In 2001, the American Medical Association revised *The Principles of Medical Ethics* for the only fourth time in the AMA's 154 year history. The "new" Section 8 now states, *A physician shall, while caring for a patient, regard responsibility to the patient as paramount.*

The addition of this language is interesting because in 1975, the AMA Principles of Medical Ethics contained the following Section 6: *A physician should not dispose of his services under terms or conditions which tend to interfere with or impair the free and complete exercise of his medical judgment.*

Today that section 6 no longer exists in *The Principles of Medical Ethics.* Why? The federal government successfully sued the AMA to remove Section 6 because it was thought to interfere with physician participation in HMOs. Also in 1975, U.S. Supreme Court ruled that the professions of law and medicine were no longer considered professions exempt from antitrust laws but were rather "ordinary purveyors of commerce" falling under the control of the Federal Trade Commission and the Department of Justice. This 1975 action formed the legal basis for today's commercialization and loss of professional status in the delivery of health care.

Capitalism and Entrepreneurs

In 1776, Adam Smith wrote, in *The Wealth of Nations*, this very apt description of capitalism. "Every individual…intends only his own gain, and he is in this, as in many other cases, led by any invisible hand to promote an end which was no part of his intention. Nor is it always the worse for the society that it was no part of it. By pursuing his own interest he frequently promotes that of the society more effectively than when he really intends to promote it."

Smith's comments are a startlingly accurate description of U.S. health care providers who both intend their own gain and are persuaded that society will benefit. The truth is—while relying on Adam Smith's invisible hand to achieve societal good is a worthy hope—our health care providers acting as capitalists are failing to deliver the universal health care that Americans want and need.

Smith's invisible hand is not a planning tool for health care reform.

Let's take a little side trip and talk about American capitalism which is, in many ways, a remarkable success. But no one claims that its benefits are available to all equally. In fact, capitalism is designed to reward some people more than others. The people who are smart, learn about the system and work hard are supposed to get ahead of layabouts and idlers. But not all disadvantaged people are loafers or lazy. Many deserve and need help.

When a benefit is designed for everyone—sometimes called an entitlement—inevitably, someone cries *socialism!* Actually, socialism is defined as a system in which the central government owns and operates the means of production.

Socialized medicine is actually practiced right here in the United States. Just look at the Army, Navy, Air Force and Veterans Administration hospitals. Care is paid for and delivered by the federal government. That does not seem to distress the senators and presidents who cheerfully get their personal health care at the National Naval Medical Center in

Bethesda, Maryland. And the U.S. war-wounded are rushed from the battlefield to receive top-notch treatment at our military medical facilities abroad.

(It is worth noting that the European countries with socialized medicine have better longevity figures than we do—while spending less per person on health care than we do.)

Now the universal health care plan proposed in this book (which you'll discover in Part 2) is not socialism but actually relies on private effort, not government ownership or operation.

However, for the sake of discussion, let's assume that health care for everyone is socialism. For those who raise that objection, the underlying implication is that socialism poses such evil that it is OK, in fact preferable, for not everyone to have health care coverage. Who decides which groups are left out? Well, that is our present situation.

The Institute of Medicine estimated that 18,000 people die each year because of no health care coverage. Cox News Service, quoting from *The American Journal of Public Health*, reported on October 31, 2007, that nearly 1.8 million U.S. veterans had no health insurance in 2004. More than half of all veterans said they had no place to go when they were sick. In addition, those veterans' households contained 3.8 million people who face a similar problem.

Some prominent people suggest that an emergency room would be the place to go for free care, paid for by those who pay. But as Mitt Romney said, "A free ride is not libertarian." And covering everyone is not socialism.

Let's examine the difference between a benefit for everyone and socialism.

Early in our nation's history, in 1786 to be precise, Ben Franklin invented the fire fighting company. Franklin's company sold memberships to subscribers who received a large medallion to place on their houses. If a house on fire displayed a medallion, the fire company arrived and doused the flames. A house afire without a medallion was allowed to burn. The perils of letting houses burn soon became clear and so, local fire departments were developed for the benefit of everyone. Denying firefighting coverage to those who did not pay was eventually interpreted as a threat to the entire community. The service was extended to all, because it was a benefit to all.

Similarly, it is an American value to offer basic education to all children, under the control of local school boards, because the entire community and the nation at large benefit by having a well-educated populace. It is democracy.

Another example is voting, considered a universal benefit of citizenship and not socialism. It is democracy.

"We hold these truths to be self-evident. All men are created equal and endowed by the Creator with certain inalienable rights: life, liberty and the pursuit of happiness." Hardly socialism!

These examples fit neither the definition nor the dynamic of capitalism. Nor are they socialism, just because everyone has a benefit.

While we are discussing capitalism and entrepreneurism, let's examine some appropriate health care data. United States spends $7,000 per person per year on health care. Other countries spend less and have better health numbers than we in the United States. (See Chapter 11 for surprising details about ten countries' health care and life expectancy.) The U.S. ranks well below other countries in longevity and neonatal death rates. These facts lead one to suspect that entrepreneurism—the quest for profits—prevails over professionalism in U.S. health care, as compared to other countries.

Here's more food for thought about our medical providers. In 1833, William Forster Lloyd lectured at Oxford University on The Tragedy of Freedom in the Commons. He said that the commons—land which is held in common by all local citizens--was vulnerable to aggressive overuse which then guaranteed mutual ruin. When everyone has unlimited use of the commons, Lloyd said, a herdsman might think, "If I put another cow on the commons, I will get all of the benefit. If there is any harm to the commons, it will be shared by all herdsmen who graze cattle on the commons." Other herdsmen follow this example with more cows until the pasture is grazed bare and can no longer feed any cows.

Garrett Hardin, Professor of Ecology at the University of California, expanded and applied the concept of the commons to many present-day economic and social problems. While he did not specifically include health care, the Tragedy of Freedom in the Commons dynamic relates perfectly to our discussion.

Health care providers offer a service, just as Adam Smith postulated; these providers in our capitalistic society rely on the invisible hand to benefit society. These providers are paid by various societal mechanisms which are the equivalent of the commons. Each provider knows that billing for additional units of service will benefit him alone but the cost of paying for these units of service will be divided and shared by society, also known as you and me. The mushrooming annual amount spent for health care is the result.

So, how can we break up the aggressive use of the commons with its guaranteed mutual destruction? Garrett Hardin offers us two ways:

1 Let government manage the commons fairly, for the benefit of all. For example: "This commons can take only 50 cows, no more."

2 Privatize and break up the commons into private parcels, each one individually owned and operated.

Professor Hardin contends that either approach can work or fail, depending on the details. "A managed commons, though it may have other defects, is not automatically subject to the tragic fate of the unmanaged commons," Hardin observes. "With real estate and other material goods, the alternative we have chosen is the institution of private property coupled with legal inheritance. Is this system perfectly just? ... An alternative to the commons need not be perfectly just to be preferable."

Let's return to health care acting as the commons, and consider the difficulties a lone practitioner faces if he/she attempts to reduce medical costs in order to benefit society. Severe financial hardship and possible closure of the practice will result.

Today's fee-for-service is not the only health care payment scheme tried in the U.S. We have also tried capitation, which is paying a physician a certain amount for each patient in the practice. The physician then was responsible for paying for all tests, lab work and other services ordered. The result was that patients received less care, lamentably, at times, less care than they needed.

It's an intuitive truth: patients need what they need (if we can determine that), not more (fee-for-service) and not less (capitation).

In real daily practice, a salary offers a physician an excellent incentive to simplify diagnosis and treatment and get it right the first time. When one is not paid for each service performed, one is not motivated to order too much or too little or to extend the therapy in any way. Systems that reward other than correct levels of treatment are not useful to the patients.

Today a promising new development is called pay-for-performance. Its purpose is to eliminate pay for units of work and to pay instead for results and outcomes. This includes disease management programs and preventive services. We'll talk more about this later.

Hospitals

Current hospital practice is plagued by a systematic deficiency. Under guidelines designed by the insurance companies to cut costs, the patient is released from the hospital as quickly as possible, which may be a good thing. But the patient may need still more than nursing home custodial care, which is frequently all that is available. What's needed but currently doesn't exist is an additional intermediate level of care.

Hospitals face other challenges. Several initiatives spell out the wide variations among hospitals in adhering to some simple best practices, such as administering aspirin immediately to heart attack victims.

One health care oversight group, called the Leapfrog Group, has asked hospitals to voluntarily report their successful use or failure to use several widely publicized best practices. The results are alarming and indicate plenty of room for improvement. For example, of the total number of hypertension patients, only about 50 percent are brought to full control. Also, aspirin is not yet given to all acute heart attack (myocardial infarction) patients.

Out Patient Clinics

In-store medical clinics are becoming a growing trend. Major retail pharmacies are providing health services inside their stores. Walgreens, CVS, Eckerd's (now Rite-Aid), Target and Wal-Mart are all moving ahead with plans to open clinics in their stores. So far they've achieved about 90 percent customer satisfaction with their limited menu of offerings.

Player 3: Insurance Companies

The insurance function is necessary. It allows the many to pay for the few. Insurance companies manage risk. They know how to skillfully manage the financial and mathematical aspects of delivering health care to large numbers of people.

Let's look at how the health insurance industry has come to this point in the United States' evolution of health care.

• Before 1929 all physicians (and as late as 1948, some physicians) simply charged according to the circumstances of the patient and the outcome. It was all according to standard medical history texts.

- In 1923, Dallas independent schoolteachers contracted with Baylor Medical Center, paying $6.00 a year per person for hospital care. This began Blue Cross. Later, not-for-profit community rated premiums were developed with open admission once a year.
- In 1929, Blue Shield was developed to cover physicians' services.
- During World War II, 1940–1949, many workers were drafted for military duty overseas, and wages and prices were frozen. In order to obtain enough workers, employers began offering employees medical care contracts as a benefit. Unions then began to view this benefit as a bargaining chip to secure better conditions for their members.
- From these medical care contracts purchased by employers, brokers, agents and insurance companies evolved. Brokers, agents and companies were privately owned enterprises.
- The for-profit insurance companies quickly offered lower premiums for young, healthy people, leaving care of the old and the sick to the not-for-profit Blue Cross and Blue Shield. This quickly put an end to annual open admission and premiums based on community ratings.
- The insurance industry persuaded Congress to create two tax advantages. The first was for employers who offer health care insurance. The second was for recipients who were allowed to receive the benefit tax free.
- In 1950, the federal government began paying for public assistance patients.
- In 1965, Lyndon Johnson started Medicare and Medicaid. Money became a factor in medicine. Hospitals raised rates and the government paid. Insurance companies charged and employers paid higher rates for their employees' coverage. This triggered increased specialization by more physicians.
- Following this, the federal government demanded that it get the lowest prices charged by medical vendors. This ended discretionary pricing.
- In 1965, Columbia Health Care Association became the first for-profit hospital.

- In 1972, Minnesota passed legislation authorizing health maintenance organizations (HMO). Since then, the indemnity policies have dwindled and managed care has increased. Today indemnity policies represent only 6 percent in a market like Kansas City. All other health coverage is managed care under HMOs and PPOs (Preferred Provider Organization).
- Dr. Denton Cooley of the Texas Heart Institute was the first to offer heart surgery for a standard fixed rate before the service was rendered. This meant the price quoted before surgery was the price charged, based on Dr. Cooley's extensive experience. Knowing the price ahead of time enabled people living in other countries as well as the U.S. to plan for the expensive surgery.
- In 1984 the federal government introduced prospective payments and Diagnostic Related Groups, or DRGs. DRGs were bundles of code numbers given to the procedures necessary to manage one diagnosis. Once these codes had been worked out, they were given to hospitals to account for the care received by patients with that diagnosis.
- To cut costs, HMOs (Health Maintenance Organizations) vigorously attempted to shift inpatient hospital care to outpatient care whenever possible.
- In 1992, President William Clinton promoted a plan to nationalize health care. This plan met vigorous opposition and failed to pass in either house of Congress. Significantly, neither political party used the opportunity to propose and consider alternative plans.

Today, insurance companies face some very heavy problems—adverse selection, cost shifting and cost control.

Adverse Selection

Let's first examine the biggest problem, called adverse selection. When an insurer has to handle more than the average expected number of expensive cases, that is adverse selection. To compensate, insurance companies raise prices which, in turn, force more people to decline health care insurance.

Here's another way to understand adverse selection.

Suppose we line up all the people with health care coverage in 2005 according to how much monetary benefit they got from their health insurance policy. Next, we divide them into deciles, or groups of 10 percent each. See Figure 1.

The first, lowest 10 percent received no money; that is to say they had no illness and filed no claim.

The last, highest 10 percent received 61 percent of all the monetary insurance benefits.

The next highest 10 percent received 21 percent of all monetary insurance benefits.

So in 2005, the highest 20 percent of insured customers used 82 percent of the health care monetary benefits. The remaining and lowest 80 percent of insured customers consumed only 18 percent of health care monetary insurance benefits.

As you can guess, the insured are well studied and understood by company actuaries. Actuaries are the leading professionals in finding ways to manage risk. It takes a combination of strong analytical skills, business knowledge and understanding of human behavior to design and manage programs that control risk. Actuaries can define accurately the financial meaning of the many who pay for the few.

It's clear why insurance companies want to have the lowest 80 percent as their customers. The problems and most of the monetary insurance

Figure 1. *People with health care insurance, divided into 10 groups. The bars represent the dollars received by each group of insured. –2005 U.S. figures.dollars received by each group of insured. –2005 U.S. figures.*

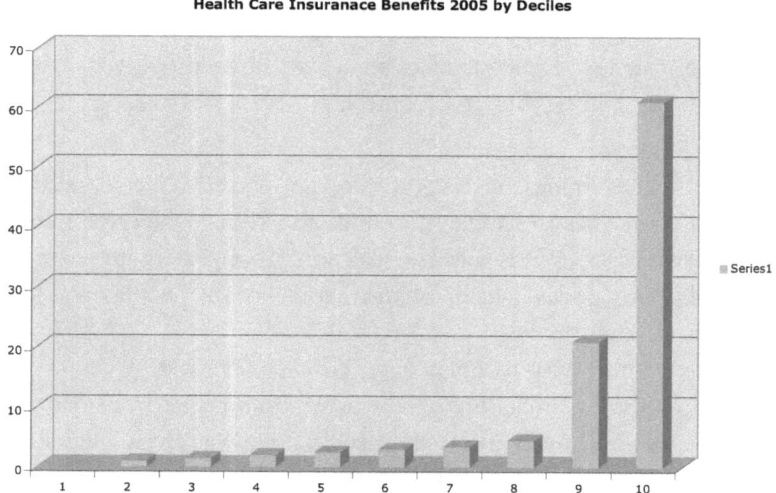

Health Care Insuranace Benefits 2005 by Deciles

benefits paid out are linked to the highest 20 percent. It is only fair to note that this distribution of benefits is expected and anticipated based on no adverse selection. It would be financially devastating to have high payout rates in 30 or 40 percent of enrollees, compared to 20 percent.

GetWell Health Care Insurance Company

To see how this really works, let's imagine that you and I and our friends put all our money together and form the GetWell Health Care Insurance Company. If we sell GetWell insurance to healthy people, they will get sick in ways that an actuary can predict and we can handle that. It would not take us long to realize that we will go broke if we sell insurance to currently sick people who have more than the expected, manageable number of expensive claims. We cannot handle people who are already sick and expensive, or about to be. That adverse selection is our biggest problem at GetWell.

What can GetWell do to make a profit, and avoid the sick, expensive people (that is, avoid adverse selection)?

- **No open enrollment!** You will never see GetWell or even a competitor display signs reading: *We will give anybody health care insurance any time they want it.* Or this sign: *Buy a cheap policy now, while you're healthy, and when you get sick, we will give you a better policy to pay your expensive bills.* Or this sign: *If you have an illness, let us pay your bills.*
- **Hire underwriters!** We will submit all applications to GetWell's underwriters who will tell us whether to accept or reject on the basis of current or anticipated expense. Insurance underwriters evaluate the risk and exposures of the prospective clients. They decide how much coverage the client should receive, how much they should pay for it, or whether to even accept the risk and insure them. Underwriting involves measuring risk exposure and determining the premium that needs to be charged to insure that risk. The function of the underwriter is to acquire—or to "write"— business that will make the insurance company money, and to protect the company's book of business from risks that they feel will make a loss. More than 1 million people, including underwriters and others working on health care contracts in America now prevent issuing insurance policies for people with expensive pre-existing conditions.

- **Sell to employers!** Employers presumably have healthy people, standing up every day, coming to work. That represents a risk pool—meaning that GetWell willingly takes the risk that these employed people are healthy. However, GetWell can refuse to put all risk pools together so that we can manage each risk pool separately. If one risk pool becomes expensive, GetWell can raise the rates at the next renewal.
- **Deny or reduce claims.** On a recent *Oprah!* TV show, the subject was health care. One of the guests was Karen Ignani, who heads the leading HMO trade association in Washington. She proudly said that the insurance industry processed and paid 5 billion claims in 2006 and denied only three percent. My pencil tells me 1,500,000 claims were denied. And it was not specified whether those claims were in the expensive 20 percent.

Clearly, adverse selection as illustrated above and as avoided by health insurers today has given us some very big problems in our current system. It's a poor system that does its best to avoid being responsible for the sickest people while making profits by doing so. Not only that. Avoiding care for sick people is a poor template for a system whose mission is to care for health.

The Uninsured

You will notice that the first scenario referred to those people with health insurance. Now let's examine the 47 million uninsured. Very likely, some undetermined percentage of these 47 million people has applied and was turned down because of pre-existing, expensive conditions.

It is reasonable to assume that the uninsured are not healthier than their insured brethren. If so, the same or worse adverse selection 80/20 proportions apply to the uninsured and represent an obstacle for insurance companies to voluntarily cover the uninsured.

Suppose we want to calculate the cheapest way to pay for 100 percent of the health care insurance costs for 100 percent of our American citizens. It is clear that if each group of 10 percent pays 10 percent of the total cost, 100 percent of the cost will be covered and 100 percent of the population will be assured of health care, even the most expensive 20 percent. This is called community rating.

Community rating worked very well when Blue Cross and Blue Shield first began offering health care insurance. It quit working when competing insurance companies cherry-picked, offering less expensive choices for young, healthy people and leaving the sick expensive people for Blue Cross and Blue Shield.

Have you heard it argued that insurance markets should be set up so that people could choose the kind of health insurance that best meets their needs and their pocket books? But think about this. Say you choose a less expensive plan now because you are young and healthy. Will an insurance company promise that when you need more benefits later, you can have them and at a reasonable cost? And, does your insurer promise that your choice will be honored now and any time later, regardless of your health?

Based upon our discussion about adverse selection, we know these better benefit choices would not be and could not be offered to the consumers when they become sick, expensive applicants.

In reality, if you allow or encourage people to choose less coverage that costs less than their percent fair share, the result will be higher costs for the remaining enrollees. Ultimately this means higher costs for everybody because everyone who survives to old age is likely to need expensive care along the way.

Or to put it another way, we need everybody's money if we're going to cover everyone, including the most expensive 20 percent, at the cheapest possible universal rate.

To make the point even clearer, if we let the bottom 80 percent pay less because they use less, no group insurance money will be left to pay for the top 20 percent so they will have to pay more. That may not seem attractive to the bottom 80 percent but mathematically, the cheapest way to achieve 100 percent population coverage is for everyone to carry their fair share of the financial burden.

"Fair share" might be interpreted as community rating or by a system for younger people to pay slightly less and older people to pay slightly more. Any deviation or relaxation of this formula, so that some pay significantly less, will result in higher costs for others. I submit that very few "bargain" health care proposals include that mathematical corollary.

Another problem with allowing choices is the necessity to hire employees, at 10 percent to 15 percent overhead, to determine which level of coverage applies when health care problems arise. Medicare has such a low overhead of 3 percent because it is not necessary to determine which tier or coverage applies: everyone in Medicare is in the same tier. Medicare Plus versions of Medicare vary in services because they are privately operated and also use more government dollars per person covered than does regular Medicare.

That puts us right back where we are today, with the uninsured falling to public welfare and hospital largess, and a large part of our population being underserved in health care.

But adverse selection is just one of the problems facing insurance companies. There's also cost shifting. Why does that cause problems for insurance companies?

Cost Shifting

One problem for providers is that emergency room health care is guaranteed for emergencies and is available even for poor people. To cover these no-pay emergencies, costs must be shifted to those who have money. This results in private paying patients, insurance companies and the government subsidizing indigent care. This cost shifting tends to mask actual costs. but it definitely increases costs.

In 1992, the Prospective Payment Assessment Commission estimated that privately insured patients were being charged, on average, 28 percent more than costs. Only the hospitals with market power have the flexibility to adjust price over time.

Another kind of cost shifting in pharmaceutical drugs increases out-of-pocket expense for the American people. Pharmaceutical companies are allowed to charge United States customers a much higher rate for drugs and medicines than they charge in other countries. Other countries' governments negotiate with the drug companies for the best price. That is not allowed in the United States.

Asymmetrical Information

Asymmetrical information is another problem for insurance companies. Here's an example: A young healthy person refuses to pay for health care insurance until he/she has a clue (asymmetrical information) that a health problem might soon become important and expensive. The young adult then applies for health care insurance.

Here's a reverse example: An underwriter refuses an insurance application because he has access to private medical knowledge, such as a DNA or genetic survey confirming the excessive risk of the applicant.

In this context, it is good to to know about the Medical Information

Bureau (MIB). Started in 1902 to help protect life insurance companies from fraud, the MIB collects personal medical information that could affect the issuance of insurance. At present, about 680 insurance companies are members. The consumer, who applies for life, health or disability insurance, receives a written notice describing MIB. The consumer permits the MIB member to ask for an MIB report by signing a statement of authorization.

You are allowed, for a fee, to review your records and correct any errors. The address is: MIB, P.O. Box 105, Essex Station, Boston, MA 02112.

Or, you can call toll-free (866-692-6901) and after answering some identifying questions, you are sent a free copy of your records. Go online to: http://www.mib.com/html/request_your_record.html

Moral Hazard

The Moral Hazard concept proposes that insurance will distort behavior because people will spend other people's money more easily than their own. It is usually used to describe patient behavior which changes with insurance. Moral hazard presumes that a person will make unwarranted use of health care benefits when a third party insurer will pay the bill. But an equal presumption might be made that medical providers will be very generous with therapy and treatment when third party coverage exists.

An article in *Health Affairs* recently poked holes in the patient Moral Hazard argument by observing that very few people have the time or the desire to abuse the system. The conclusion was that patient Moral Hazard is a considerably overblown criticism. As Timothy Noah, online *Slate* columnist observed, "Frivolous visits to the doctor are comparatively rare and confined to a few hypochondriacs and hysterics. Going to the doctor is, after all, a fairly unpleasant experience." Not to mention time consuming, especially for people with responsibilities. Employees, especially part-timers, are usually very limited in their discretionary time and find it difficult and expensive to lose work time for a doctor's appointment.

Cost Control

Health care costs are rising at two times the rate of inflation. Controlling U.S. health care costs has proven very, very difficult. America has tried some process measures such as health maintenance organizations, preferred provider organizations, and capitation payments for physicians, to no avail. Cost controls on how we pay (process) must equate with what we pay for (content).

Medicare expense, since its advent in 1967, has grown 2.5 percent faster annually than the gross domestic product (GDP). Medicare was 3.1 percent of GD in 2006. If this growth rate is extrapolated to the year 2040, Medicare will represent about 20% of GDP; I think that is very unlikely to happen. But this is an alarming rate of growth in Medicare. To change this outcome will require cost reconciliation, and the method has yet to be determined.

There have been two notable attempts, by the CMS, to control costs. The first, over a number of years, was to reduce physician's payment rates. Only by heavy lobbying of Congress were these reductions set aside some years.

The second, very recently, was a decision that Medicare would not pay for hospital complications that were avoidable. The utility of this decision is unknown, because the decision is so new.

Problems that insurance companies create:

1   The number of competing health plans has shrunk; this reduces competition and increases insurers' profits. This also puts providers in an awkward position while insurers gain added leverage in setting health care standards.

2   Many physicians complain that insurers are slow to pay when claims are submitted. This is an advantage for the insurance company which profits on the float of money. However, slow payment is devastating to a practicing physician whose overhead never stops. Clearly, there is a need for effective governmental supervision of insurance companies' financial practices.

3   Insurance companies have been criticized for refusing benefits or diminishing benefits on behalf of their insured customers. Thus, the value of "insured." is uncertain, and the test of an insurer's performance is to need the help and make a claim.

PLAYER 4: Federal Government

The federal government needs to help people who cannot afford health care on their own or who are entitled to health care in a statutory program such as the Veterans Administration. The government currently pays about 47 percent of health care expenses in America.

Many people in the United States are opposed to entitlements. Universal health care coverage would have to be considered the "Mother of All Entitlements." As a result, universal health care consistently attracts opposition.

To address the anticipated unsustainable cost growth of Medicare, the government plans to cut physicians' reimbursements and withdraw billions from Medicare entitlement.

A Sustainable Growth Rate (SGR) formula reduces physician payments each year. For example, in 2007 physician payments will be reduced by 5.1 percent. Only by complaining and bargaining are physicians able to gain exceptions for some years. The truth is, payments to physicians have never exceeded 20 percent of the total federal health care expenditure in any year. If you were able to get physicians to work for free, it would not solve the health care expense problem. Unfortunately, these reductions in payments to physicians are resulting in fewer applications for residencies in the most complicated specialties.

On February 1, 2007, President George W. Bush proposed withdrawing 70 billion dollars from the Medicare budget over the next 5 years. It is a reminder that, "He who pays calls the tune."

At the same time—inexplicably, or predictably—the government decided to protect pharmaceutical companies by eliminating price negotiations for medications under Medicare Part D. A study by Dr. Alan Sager and Deborah Scolar of the Boston University School of Public Health shows that 61 percent of the federal Medicare Part D money will go to drug makers in the form of 38 percent added profits. In addition, the federal government has made it illegal to import medications from Canada where price negotiation with pharmaceutical companies has lowered drug prices significantly.

PLAYER 5: New Makers

New Makers are people who introduce new ideas, new policies, new drugs, new devices, new technologies, and new knowledge and new treatments of all kinds. It is extremely important that the health care system properly benefits from the output of new makers. The operative idea is properly benefit. That is very difficult. Uncritical acceptance of new innovations can be very expensive and unrewarding. Likewise, uncritical blocking of the new prevents proper advancement.

The United States has clearly profited from inventions, technology, new drugs, advanced biological research and ethical practices. Medicine is extraordinarily complicated and our physicians are among the best educated in the world. The United States is the center of advanced medical and surgical research and teaching. This leadership is to be valued. But it is also expensive and susceptible to overuse, corruption and exploitation.

Of course, we do not want to abandon world leadership in medicine, research, technology and teaching. The trick is to find the balance between accepting everything new and rejecting everything new. We must develop a method of judging all the new inventions and policies so that we can accept, promptly, the best and the most rewarding of the new. For example, the Mayo Clinic, a pioneer, went smoke-free in 1986. The rest of the country is belatedly and sporadically joining the smoke-free movement.

Yet another problem soon will affect our world leadership. The United States is losing its primacy in college education, according to a study conducted by the National Center for Public Policy and Higher Education. "Although the United States still leads the world in the proportion of 35 to 64-year-olds with college degrees, it ranks seventh among developing nations for 25 to 34-year-olds. This group is lagging educationally compared with the Baby Boom generation," according to the report. "Other nations have significantly improved and expanded their higher education systems. Perhaps for the first time in our history, the next generation will be less educated."

PLAYER 6: Employers

The role of the employer as the provider of health care insurance was born during World War II when wage and price controls were instituted. Employers needed to attract workers but couldn't use the usual financial incentives. They devised the practice of offering health care benefits to lure workers. One might say that capitalism triumphed in meeting a need. One might also say that business owners dismissed patriotism and subverted the goal that drove the wage and price controls.

In any case, employers were "hoist on their own petard" because unions immediately seized on health care benefits as a bargaining chip in labor contracts. The insurance companies used lobbyists to prevail upon Congress and thus health care benefits became tax-free income to workers and a tax deduction for the employer. Since World War II, U.S. employers attract workers by offering some level of health care benefit.

Employer-Based Health Care Insurance

In its heyday, 80 percent of adult health care was obtained through employer-sponsored insurance. Today, that level has slid to less than 60 percent with further erosion every year, especially among small businesses. Large and small businesses are shifting a larger percentage of health care costs and content to their employees and retirees.

In a recent accord with the United Auto Workers, General Motors offloaded $55 billion of employee and retiree health care benefits to a voluntary employee benefit association (VEBA), General Motors agreed that medical benefits for hourly workers and retirees and their families will remain in place for the next two years.

In the *Wall Street Journal*, September 25, 2006, a writer noted that as employees pay a larger percentage of their health care premiums, they ask employers to help them understand their new and many options. Providing this guidance as well as the insurance is proving costly to employers in both time and finances.

Large corporations, which spend billions of dollars on health care, have decided to purchase health care from high-quality providers and, if possible, at a reasonable rate. These corporate health care analysts use data from information technology to achieve this goal. As medical information technology develops and matures, it is likely that the designation high-quality provider will help the corporate health care purchasers while also stimulating improvements in the remaining providers.

Yet it is very difficult for a single employer to contemplate complete relief from this health care responsibility because the competition continues to offer this benefit. Ironically, employer-provided health care benefits remain, legally, a voluntary commitment. Fifty percent of small business owners say the health care benefits they can offer are the major obstacle when competing with large businesses for employees.

Then too, the employer remains a valuable customer of insurance companies because he offers a risk pool and multiple contracts. Insurance companies have no reason to oppose profitable employer based health care benefits.

Employers, like insurance companies, have the problem of adverse selection. As long as employees are healthy—read not expensive—the interests of the employer and the employees are in alignment; everything is rosy. In 2005 a Wal-Mart interior memo recommended changes that would "dissuade unhealthy people from coming to work at Wal-Mart." An ordinarily healthy employee in an automobile accident that results in ventilator therapy in ICU for a month brings up some issues. The employer's insurance rates are certain to rise markedly next year. The employee, who cannot come to work, will probably lose his job together with his employer-provided health care insurance. COBRA will allow him to purchase his health care insurance for 18 additional months, if he can afford it. This also is a poor template for a system designed to care for health.[16]

In many states, laws permit insurers to raise health premiums substantially for small employers when one worker incurs significant medical bills. State-by-state differences in insurance regulation make the problem worse.

David Fear, president of The National Association of Health Underwriters, points out, "There are 50 sets of state rules." Large companies have enough employees—a risk pool—to self-insure. Small companies do not enjoy the benefits of a large risk pool.

How many Americans lack health insurance? How does being uninsured affect access to preventive care?

In 2004, 64 million Americans—26 percent of the nonelderly population—were without health insurance for at least one month during the year, and 34 million of these individuals—14 percent of the nonelderly population—were uninsured all year long. Those without health insurance were less likely than those with coverage to receive preventive care services at appropriate ages.

See the charts on the next page for a graphic illustration of the uninsured at various timepoints, and the blood pressure and cholesterol screenings they did or did not receive.

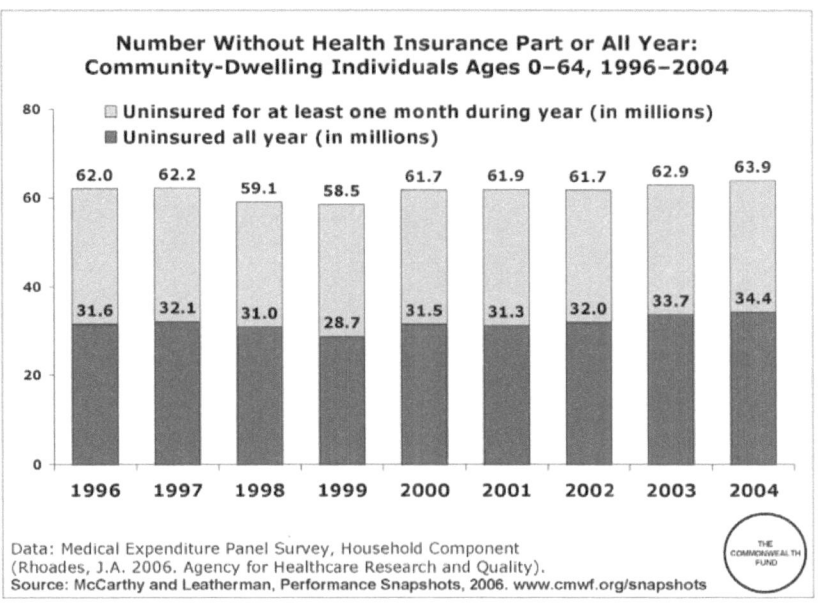

Number Without Health Insurance Part or All Year: Community-Dwelling Individuals Ages 0–64, 1996–2004

☐ Uninsured for at least one month during year (in millions)
■ Uninsured all year (in millions)

Year	Total	Uninsured all year
1996	62.0	31.6
1997	62.2	32.1
1998	59.1	31.0
1999	58.5	28.7
2000	61.7	31.5
2001	61.9	31.3
2002	61.7	32.0
2003	62.9	33.7
2004	63.9	34.4

Data: Medical Expenditure Panel Survey, Household Component
(Rhoades, J.A. 2006. Agency for Healthcare Research and Quality).
Source: McCarthy and Leatherman, Performance Snapshots, 2006. www.cmwf.org/snapshots

THE COMMONWEALTH FUND

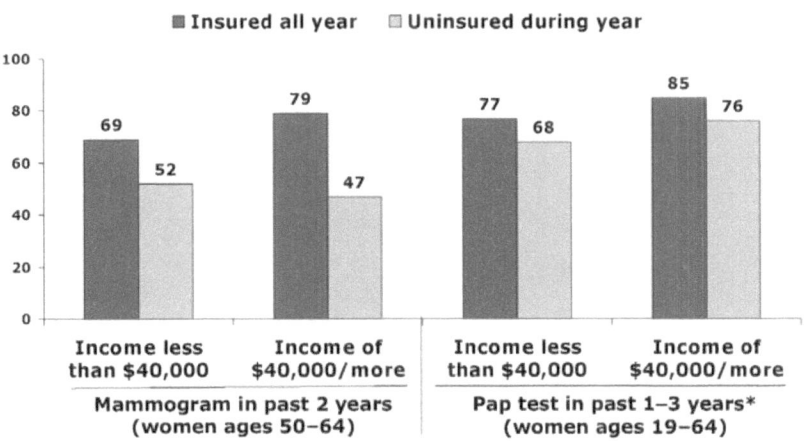

Percentage Who Received Cancer Screening: Community-Dwelling Nonelderly Adults by Insurance and Income Status, 2005

■ Insured all year   ☐ Uninsured during year

	Mammogram in past 2 years (women ages 50–64)		Pap test in past 1–3 years* (women ages 19–64)	
	Income less than $40,000	Income of $40,000/more	Income less than $40,000	Income of $40,000/more
Insured all year	69	79	77	85
Uninsured during year	52	47	68	76

Data: 2005 Commonwealth Fund Biennial Health Insurance Survey (Collins, S.R.
et al. 2006. The Commonwealth Fund). *Pap smear in the past year for women
ages 19–29 and in the past three years for women ages 30–64.
Source: McCarthy and Leatherman, Performance Snapshots, 2006. www.cmwf.org/snapshots

THE COMMONWEALTH FUND

Zeus does not bring all men's plans to fulfillment.

Homer (800 BC - 700 BC), The Iliad

CHAPTER 3

Problems

C ertain problems within our present system must be resolved. You might even call these diseases within our present incremental health care system.

The Three-legged Threat

Health care frequently grapples with the three legged threat comprised of Access, Cost and Quality or Benefits. These are intimately and hydraulically linked together. Quality affects cost and cost and quality affect access.

Accessibility of Health Care

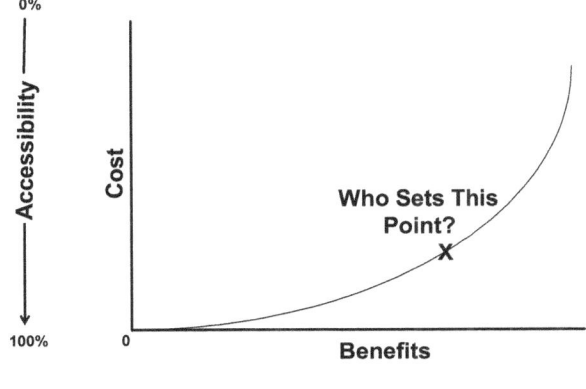

The diagram above shows the difficulty in providing 100 percent accessibility in a free, voluntary market. If you offer zero benefits, everyone can afford it. As benefits increase, so does the cost. As the cost increases the accessibility decreases. If a decent benefit package is guaranteed to everyone, there must be financial help for those who cannot afford the package.

47 Million Americans Uninsured

As discussed earlier in this book, the number of uninsured grows each year. Most of this growth is because increasing costs discourage purchase. But a good portion of this growth can be traced to lack of offer: certain people are not offered or are turned down for health insurance.

According to the Institute of Medicine, approximately 18,000 Americans die prematurely each year because they have no health insurance.

They cannot afford the insurance or they have been turned down when they apply, and thus cannot obtain treatment services when they get sick.

Admittedly, the number 47 million has its nay-sayers and yay-sayers.

For example, George Will, commentator on pro-life and family values, diminishes the number by subtracting:

11 million as illegal immigrants;

26 million as transiently uninsured;

Leaving only 10 million therefore chronically uninsured.

The opposing arguments are:

No one knows the precise number of illegal immigrants.

Illegal immigrants need health care too.

26 million are transiently uninsured at any one time-- therefore over a year's time, many more than 26 million are uninsured.

We do know that of the uninsured, 29 percent are Hispanic, 15 percent are black, and 50 percent are white non-Hispanic.

The ages of the uninsured are interesting:

Of children under 18:............... 12 percent are uninsured,

Of young adults ages 19 to 24:.... 35 percent are uninsured,

Of young Hispanic adults:......... 56 percent are uninsured.

In 2005, more than half of Americans whose yearly incomes are less than $20,000 said they were uninsured for some or all of the year.

National surveys show that the primary reason people are uninsured is the high cost of health insurance coverage.

Lack of Planning and Coordination

Perhaps Brian Klepper, Director of The Center for Practical Health Reform, says it best. "Mechanisms of American health care encourage millions of professionals and corporations to make self-interested decisions daily, independent of their effects on the larger system. Rich reimbursements, currently 1.8 trillion dollars, have made many industry groups wealthy and have translated into tremendous political influence broadly distributed among the players. Gridlock results because every proposal threatens the interests of some constituency with the influence to kill it."

The Tort System

Our present medical liability insurance system does not well serve either physicians or patients who have been injured. Both patients and physicians complain that the system is unfair.

Many injuries or costly poor results are never reported or reconciled in any way. Many, perhaps even most, medical injuries do not reach the tort system. Medical injuries that do reach the system take the form of a lawsuit. Those lawsuits sometimes are frivolous and other times are justified. In whichever case, lawsuits are frequently won by physicians at great cost of money, time and emotional stress for both plaintiff and defendant.

Those cases won by the plaintiff, or patient, may receive very large financial awards. Lawyers for both plaintiffs and defendants always do well!

The result of all this is very high malpractice insurance premiums for doctors and in some cases, the increasing extinction of certain types of practice in states with high litigation experience. The tort system clearly needs reform for the benefit of all concerned.

We Are Slow

Survey organizations such as the Leapfrog Group have shown statistically that the U.S. is slow to achieve general adoption of known best practices, such as positioning respirator patients upright in bed to lessen pneumonia, or, giving aspirin routinely to myocardial infarction (heart attack) patients, or, administering anticoagulating (anti-clotting) drugs to lessen the effects of a stroke (atrial fibrillation).

The U.S. is also slow to computerize medical records and to post on-line accountability scores for doctors and hospitals. Only 30 percent of physicians' offices in the U.S. are computerized, and even those computerized records are not readily available or accessible for accountability purposes.

In the Netherlands, 98 percent of primary care physicians' offices are computerized. In the United Kingdom, it is 89 percent.

Again, we are behind.

CHAPTER 4

Recent Efforts to Improve the System

The problems we've explored thus far have not gone unnoticed. Many have come forward to propose changes in our health care system and ways to extend coverage to everyone. Let's look changes already underway or on the table today.

Employer Plans

Some time ago, 80 percent of adults under age 65 had health care coverage by means of their employers' voluntary health care benefit package. Today, rising costs of health care benefits are causing employers—especially small employers—to drop employees' health care coverage.

Even the larger corporate employers are shifting more of the cost to employees. This cost shifting causes employees to think twice before signing up for health insurance. Less than 60 percent of adults under 65 now have health care coverage from their employers.

Another employer strategy encourages employees to take more responsibility for their own health by improving nutrition, losing weight, exercising, and stopping smoking.

Medical Savings Accounts/Health Savings Accounts

Congress enacted health savings accounts (HSAs) in 2003 as part of the Medicare reform package. The HSA is essentially a medical savings account (MSA), according to the Council for Affordable Health Insurance.

A health savings account is a tax-sheltered savings account similar to the IRA, but earmarked for medical expenses. Deposits are 100 percent tax-deductible for the self-employed; this now includes everyone with an HSA. Funds can be easily withdrawn by check or debit card to pay routine medical bills with tax-free dollars. Larger medical expenses are covered by a low-cost, high deductible health insurance policy. What is not used from the account each year stays in the account and continues to grow interest on a tax-favored basis to supplement retirement, just like an IRA.

When combined with a required, low-cost, high deductible health insurance policy, the health savings account is meant to replace a traditional high-cost health insurance policy. A health savings plan will restore a high degree of freedom of choice by allowing you to choose your own physician--typically from an extensive PPO directory—without the extensive restrictions imposed by HMO-type plans.

The savings accounts are intended to help pay smaller covered medical expenses until the deductible is met; should the need arise, the high deductible insurance policy takes care of covered medical expenses exceeding the deductible.

Consumer Directed Health Care

One interesting proposal is Consumer Directed Health care.

The essence of the program is a tax-favored, high-deductible, health savings account combined with a catastrophic health care policy. The policyholder deposits money to be used later to pay the high dollar deductible. If the money is not used, it is saved tax-free. With good luck, the account could accumulate enough money to pay for some sizable health care bills, with the catastrophic policy covering the largest expenses.

Advantages:

1   The plan is said to increase individual responsibility and budgetary decision-making.
2   Savings are immediate due to the lower premium for the catastrophic coverage policy.
3   It lets the consumer decide on the amount to be saved for the

deductible and therefore on the amount of saving in the premium of the catastrophic coverage.

4 It's attractive to a great many people.

Disadvantages:

1 Combining a health care reform with a tax savings program weakens both.
2 The tax savings feature does not apply equally across all income levels. It benefits the high-tax population but not the population that pays no tax.
3 The health care reform is also beneficial to the high income population. The low income population has difficulty covering enough deductible to achieve the savings in the catastrophic premium.
4 Any plan that reduces the financial contribution of the population also reduces the possibility of covering the expenses of the upper, expensive 20 percent.
5 The system's advantages are chiefly confined to acute-care. Chronic disease management will regularly consume the high deductible at great expense. This is easier for the wealthy than the poor.
6 Will insurance companies be able to expand insurance coverage in the uninsured group?

SCHIP

This is an acronym for State Children's Health Insurance Program, which is administered by the states. However, the program is federally mandated; the federal program is called the Children's Health Insurance Program (CHIP)

This federal program is designed for families who earn too much money to qualify for Medicaid, yet who cannot afford to buy private insurance for their children. CHIP coverage provides eligible children with coverage for a full range of health services, including regular checkups, immunizations, prescription drugs, lab tests, X-rays, hospital visits, and more.

President George W. Bush's Plan

President Bush's health care proposal would use tax breaks to make it easier for people who do not have employer-provided health insurance to buy coverage on their own. The tax incentives would be similar to deductions used by homeowners for the interest on their mortgages. The current health system relies primarily on employers to provide health care coverage as a fringe benefit which is not taxed. The Bush plan would set a cap on the amount of coverage that could be offered tax-free. Anything above that would be taxed as income.

President Bush says it is a $41 billion "prosperity" plan—a $35 billion tax credit which could help up to 18 million Americans buy private health coverage.

Bush's proposal would provide a write-off of up to $2,000 for families earning $30,000 or less, or $1,000 for individuals earning $15,000 or less. Those who pay no taxes would get a check when premiums come due.

Part D Medicare Pharmaceutical Plan

More than 10 million people have enrolled in a stand-alone prescription drug plan that is part of Medicare and for the Medicare population. The price of drugs is higher for the government under this Medicare plan than was the price of drugs under Medicaid.[20]

Missouri (and other states) Plan to Eliminate Medicaid

In 2005, the state of Missouri cut Medicaid because of budget problems. More than 100,000 Missourians lost Medicaid coverage. Some 300,000 people experienced a reduction in services. The state relinquished millions of federal Medicaid support dollars. The Medicaid program is due to end June 30, 2008.

CHAPTER 5

The Oxymoron

By now, you may have concluded that offering universal health care for all within our incremental health care system simply will not work. Frankly, I find that health care for all and incremental health care are oxymorononic. The two concepts are polar-opposites and cancel each other out.

So what have we learned?

1. We are ideally set up to achieve the health care results we are now getting.
2. The imminent collapse of the system is no surprise. The problems we face have piled up over many years and all the players have contributed.
3. Employees' health care is not the legitimate business of employers.
4. Insurers seek to please the employer who is their customer— not the person being insured.
5. Employer insurance offering is voluntary. Insurance participation by employees is voluntary. It is impossible to achieve universal health care in an incremental system which allows voluntary participation for both provider and receiver.
6. Time is of the essence. We must take action now to help our inadequate health care system.

I believe that our present, United States' incremental, voluntary, employer based health care system is and always will be dysfunctional and disappointing.

Now, join me in Part 2, to see that we are not all that far from a solution—one that I think will work for everyone.

The pessimist complains about the wind;
The optimist expects it to change;
And the realist adjusts the sails.

William Arthur Ward

PART 2

The Health Care We Need

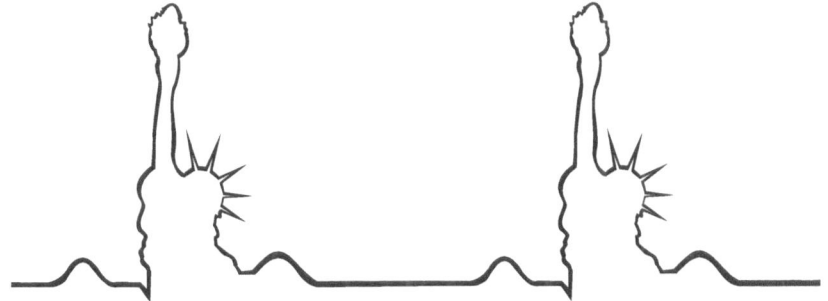

We now realize, as we never realized before,

our interdependence on each other;

that we cannot merely take but we must give as well;

that if we are to go forward,

we must move as a trained and loyal army

willing to sacrifice

for the good of a common discipline,

because without such discipline

no progress is made, no leadership becomes effective.

Franklin Delano Roosevelt, 1933

CHAPTER 6

The Big Picture

M
any highly intelligent people are working on health care reform. They seem to divide themselves into two distinct groups. The first group is trying to improve one or more increments of care in our incremental system. They are trying to improve the mess and to cover more of the uninsured. So far no one in this group has managed to create a plan that fixes everything. It is my conclusion that success for this group is impossible.

The second group is trying to cover 100 percent of Americans and, in the process, fix the big problems we have discussed. I am in this group. It is interesting that the two groups rarely see any way to deal with each other—it's almost as if a wall separates them.

The wall is getting thicker as accusations heat up. One presidential candidate hurled the accusation of "socialism" recently at universal health care plans advocated by his opponents. He said those who urge "mandatory" universal health care were "moving toward socialized medicine so fast, it'll make your head spin."

This candidate is not alone in his belief. My response is simple: The American people overwhelming have said they want health care for everybody.

When attempts to plan for universal health coverage are disparaged as "socialism," I must conclude that the critic does not want all Americans to have health care as a right of citizenship. Is health care to remain only for the fortunate in America?

Through the discussion in Part 1, I hope you can clearly see that our present health care system—a voluntary, incremental and employer based system—is not only not now working for everyone, it cannot work to cover everyone. Our dysfunctional health care concepts need to be changed to provide health coverage and better health for everyone in America, at a cost that won't bankrupt us.

It's time to provide a window onto a new way of looking at health care—a glimpse into the future—a different way of doing health care.

We'll explore the principles driving good health care. It is important to understand them so that we all recognize, appreciate, and use these principles to evaluate and judge proposals for health care change that reformers are offering to America.

Good health is the goal. If everyone stayed healthy, we would not need health care. But not everyone is or stays healthy. We have to rely on our health care system to deal with that. How should we proceed?

Three Proposals for Change

Uwe Rheinhardt, professor of Health Care Policy at Princeton University, asks the crucial question: "Are we ready to treat health care like we treat education?" Both education and health care are, after all, accepted as necessities for our own societal self preservation.

The year 2006 was a watershed because three important health care reports were issued. One report was by average citizens, another by a health care think tank, and the third report by respected health care professionals. All three reports, from these very different sources, recommended the same, or very similar, overlapping changes to our health care system, as you will see.

The three organizations are—
- The Citizens Health Care Working Group
- The Commonwealth Fund Commission
- The Mayo Clinic National Symposium on Health Care Reform.

Using their recommendations as an outline, we'll set up a list of changes and then divide them according to who must accomplish them. We will compare the groups' recommendations against the Plan we propose. First, let's look more closely at who made up each group and what they said.

The Citizens Health Care Working Group

This group came out of the Medicare Prescription Improvement and Modernization Act, passed by Congress in 2003. Among other things, this bill required the federal government to interview U.S. citizens about their wishes for a health care system. The Citizens Health Care

Working Group was created. In October 2006, after three years of meetings held around the country, the group issued a final report summarizing citizen input. American citizens, the report says, want the government to:

1   Establish public policy that all Americans have affordable health care
2   Guarantee financial protection against very high health care costs
3   Foster innovative integrated community health networks
4   Define core benefits and services for all Americans
5   Promote efforts to improve quality of care and efficiency
6   Fundamentally restructure the way end of life services are financed and provided.

The Commonwealth Fund Commission on a High-Performance Health System

The Commonwealth Fund is a private foundation and think tank, founded by Anna Harkness in 1918. This foundation is devoted to improving health care in the U.S. In 2006, the Commonwealth Fund recommended the following core values and goals:

1   Expand health insurance coverage to all residents
2   Implement major quality and safety improvements
3   Work toward a more organized delivery system that emphasizes patient-centered primary and preventive care.
4   Increase transparency and reporting on all quality and costs
5   Expand the use of interoperable information technology
6   Reward performance for quality and efficiency
7   Encourage public-private collaboration to simplify and create more effective change.

The Mayo Clinic National Symposium on Health Care Reform

This May 2006 meeting at the Mayo Clinic was attended by more than 200 national leaders from business, health care, government, public policy and patient advocacy. From this gathering, these leaders issued a public report recommending that the U.S.:

1   Build a mandate for change.
2   Reimburse health care based on results
3   Encourage formation of integrated systems
4   Increase transparency—information sharing—among systems

and physician practices
5   Define essential health care services
6   Reward patients who select high-quality providers.

As you've noticed, these reports make very similar recommendations which actually form a kind of consensus. We'll refer to this as the **blue ribbon consensus** as we explore health care in the U.S. and how to improve it for everybody.

We can further divide these recommendations into content--what is to be offered--and process--how it should be offered and paid for.

If these recommendations were easy they probably would have been in place by now. As we learned in Part 1, the current system is well-established and has allowed medicine and health care to deteriorate over many years. If only sweet reason was used, we would already see a different and improved health care system with vastly different performance. But many factors work against such change. Realistically, it will take time to analyze and improve performance, costs and outcome, or content.

Mature information technology is the key to long-term improvement in cost control, performance and efficiency, no matter what health care system that we ultimately live with.

We could, however, make some immediate improvements to the system. Let's look!

The best interest of the patient is

the only interest to be considered.

William J. Mayo, M.D.

CHAPTER 7

The Plan

You've already noticed that the blue ribbon consensus, formed by the three groups, is unanimous in calling for basic health care that covers everyone in the United States. Here's how each group said it:

- Define core benefits and services for all Americans (Citizen group)
- Expand health insurance coverage to all residents.(Commonwealth Fund)
- Define essential health care services (Mayo Clinic symposium).

The Plan we propose

We propose, for the United States, **a system of mandated, privately owned, basic health care policies that provide universal coverage.** The acronym for that is PMIUC, which stands for private, mandated, insured universal coverage.

This Plan incorporates the blue ribbon recommendations, and goes a bit farther. It is a privately owned plan that includes six of the seven current players in the present health care system.

Seven Rules Operate the Proposed Health Care Plan

In order to work, this Plan has seven rules which define it and enable it to work. The first six rules are easy to understand. Rule number seven is a little more unfamiliar to present day thinking. But all seven rules are needed and really are not such a stretch to set up. Let's look at these rules, one at a time.

RULE NUMBER ONE: Establish a large central computer system

This centralized system would be bought and paid for by the insurance companies. It would function as a central insurance system for the insurance companies like the Federal Reserve System is the central bank for the U.S. Like the Federal Reserve, the central insurance system would be privately owned, but chartered by the federal government to perform mandated duties.

One duty could be to generate ZIP code modifiers that help set policy fees. These modifiers are important because it's cheaper to offer health care in Des Moines, Iowa, than it is in New York City.

Its information technology would generate public reports about quality and results.

This central computer and its information technology system would also organize and analyze cost and clinical information from providers. Such analysis would help identify successes in best practices, efficiency, management of chronic disease, and outcomes of acute and chronic care. Ultimately, the central insurance system might facilitate a central depository for patient medical records.

RULE NUMBER TWO: One basic health care plan serves everyone

The concept of basic coverage is fundamental to the success of any plan for universal health care. No matter what reform is undertaken, it must define what it will offer. By some as yet undefined method, basic coverage must be spelled out, similar to the way that Medicare is a basic policy. Here's one recent proposal for basic coverage—we'll identify the source later:

- All American citizens receive benefits
- Primary and preventive care
- Inpatient and outpatient services
- ER services
- Prescription drugs
- Long-term care
- Mental health services
- Dental coverage
- Hearing and vision coverage
- Substance abuse treatments
- Chiropractic services
- No deductibles, no co pays, no out-of-pocket expenses
- Portable, see any doctor in United States
- Preventive services included.

FIGURE 3	Preventive services that would provide the most health benefits to adults if utilization rates improved.*	
Services	**Percentage of Adults Currently Receiving Services Nationwide**	**Additional QALYs[b] Saved if Services Extended to 90% of Adults Nationwide**
Tobacco use screening and brief intervention	35%	1,300,000
Colorectal cancer screening	35%	310,000
Influenza vaccine in adults	36%[c] in ages 50-64 65% in ages 65 +	111,000
Breast cancer screening	68%	91,000
Cervical cancer screening	79%	29,000
Chlamydia screening	40%	19,000
Pneumococcal vaccine in adults	56%	16,000
Cholesterol screening	87%	12,000
Hypertension screening	90%	0
Based on limited available data, utilization rates of 50% were assigned to the following services:		
Aspirin chemoprophylaxis	50%	590,000
Problem drinking screening and brief counseling	50%	71,000
Vision screening in adults	50%	31,000

*Maciosek MV, Coffield AB, Edwards NM, Goodman MJ, Flottemesch TJ, Solberg LI. Priorities among effective clinical preventive services: results of a systematic review and analysis. *Am J Prev Med* 2006.
[b]**QALYs** = Quality adjusted life years

What preventive services would bring the most "bang for the buck?" See Figure 3, previous page, for the Partnership for Prevention's proposal. Note that in the last column is estimated number of years with improved quality of life (QALYs) that could be enjoyed if these services were received by 90% of all adults, nationwide.

RULE NUMBER THREE: Every insurance company must offer the basic policy

- Insurance companies can only compete on cost and quality, not by changing the basic policy benefits. Insurance companies may, if they choose, offer supplementary insurance subject to underwriting. Any supplementary policy is optional for the insurer and the insured who is always covered by the basic policy.
- This saves overhead. Insurance companies would no longer need to ascertain levels of contract coverage when claims are made. They would just process the claim.

RULE NUMBER FOUR: Insurance companies are required to accept all applicants for the basic policy

No pre-existing conditions. People who need health care are covered at once. Since we know that the uninsured use about half of what the insured use and 15 percent of population is uninsured, we can deduce that insurance for all would represent about 7 ½ percent rise in demand and cost.

RULE NUMBER FIVE: Everyone must buy the basic policy; those with lower incomes will receive federal government financial help

To achieve the lowest premium price for everyone, mathematics requires that everyone purchase the basic policy. Each person who does not buy the basic policy decreases the available pool of money and increases the cost for those who do purchase the policy. So, everyone must buy the basic policy.

To make sure everyone signs up, each citizen submits a record of coverage with the annual income tax return. If coverage was not obtained for the full year, the taxpayer is liable for a tax penalty which could go into a fund for the uninsured. This rule helps the insurance companies, to some extent, with adverse selection.

Financial assistance will be available for those at the bottom of the income ladder. Based on the experience of European countries, we could project that about 30 percent of the U.S. population would need varying amounts of financial help. In total, the financial help would amount to about 20 percent of the budget.

Mandated or required purchase of the basic health care policy is objectionable to some people. However, it is necessary for four reasons:

1 To cover everyone
2 To reduce overhead by eliminating choice
3 To partially protect insurance companies from adverse selection
4 To maximize available funds for medical benefits.

RULE NUMBER SIX: Each person has the right to switch insurance companies if they find better service or better results elsewhere

- Every American has the right to select one insurance company from all those available nationwide. All insurers will offer the same basic policy but prices, service and claims handling will still vary.
- So, the policy holder may change to another insurance company at any time. If not well served, he or she may switch to another insurer with a better track record. Websites which rate insurers against various chronic illness and other claims will undoubtedly spring up to help the process along. The customer, who after all is paying for the policy, is encouraged to seek the best, most cost effective service and health care.
- This puts insurance companies on notice that poor performance —slow processing, disputing payments, or unpleasant customer service—will rapidly deplete their customer list.

RULE NUMBER SEVEN: A National Risk Pool with Incentive Reinsurance is established

This rule is only possible with the other six rules in place. A national risk pool combined with a plan for incentive reinsurance are two extremely important concepts. I believe that mandated, individually owned health care policies can be successful only with the help of these concepts.

Health care insurers that compete effectively and successfully by offering excellent service and quality might find themselves flooded with sick people who flock to them because of their good record. Even with everyone's financial participation, excellent work could eventually adversely affect the performing insurance company's profitability. Let's remember that insurance companies are not legally obligated to offer health care insurance. If things are not going well, they can call it quits and leave the health care field entirely.

Incentive reinsurance and a national risk pool would serve to reward and level the playing field for successful, service-oriented insurers.

Here is how it would work.

- At the beginning of every month, insurance companies send a percentage—say, 20 percent—of their basic policy revenues to the central computer system (see Rule Number One). The money is placed in the national risk pool.

- At the end of every month, insurance companies send in their "experience codes," which are like the present medical codes used by all physicians and hospitals when submitting insurance claims. These experience codes identify the kinds of claims paid out by each insurer.

- Based on the severity of their experience codes, insurers are eligible to draw back money from the national risk pool.

- If all companies have the same performance profile, they will all get their money back from the national risk pool. But companies that lift heavier loads, so to speak, by caring for sicker patients (as indicated by experience codes) will receive more money back from the national risk pool. This is **incentive reinsurance**.

Incentive reinsurance with a national risk pool has several advantages.

First, if a pestilence occurs, such as a SARS outbreak in San Francisco, for example, insurers who took excellent care of the SARS victims might suffer a severe financial hit. However, their heavy experience codes would entitle them to draw back money in compensation from the national risk pool.

Second, an insurance company might decide to intentionally repel expensive customers, such as diabetics, thus relieving themselves of that expense. But the company's "lighter" experience codes lessen the drawing of funds from the national risk pool. On the other hand, the insurer that took care of the diabetics will be compensated from the national risk pool.

By adjusting the percentage of premium withheld and weighting the services to be encouraged, this construct becomes ideal incentive reinsurance. It provides incentives for insurers to control costs as well as to attract and service difficult cases—exactly what we want in a health care system. Insurance companies could actually lose money by not serving the difficult-to-treat diseases.

Hybrid: Both Market-Driven and Mandated

So, the Plan described within these seven rules combines features of both a free, market-driven system and a controlled, government mandated system.

The Plan is a **market-driven** health care system to the extent that health care insurance companies remain private and will compete on cost and quality.. It also introduces an incentive for health care insurance company to improve their profits by improving preventive measures, performance and outcomes. Health care insurance companies will still set their own premium prices. However, under the Plan, health care insurance companies can gain market share by doing things cheaper or better or both. Their customers in the American public will be armed with information about each company and thus be better able to shop policies based on price, results and coverage offered.

The Plan is a **mandated** system to the extent that the government specifically regulates the parameters—the basic package—within which health care insurance companies must operate. Under this Plan, health care insurance companies no longer have the right to refuse coverage for any reason including pre-existing conditions.

The Plan does protect health care insurance companies from adverse selection by the "federal reserve system-type" reinsurance program and a reinsurance pool funded by the insurers themselves.

The Plan does not allow health care insurance companies to benefit any longer from unrestricted profit decisions. Instead, it serves the public by relying on actuaries, the power of information technology, and mathematics.

At the end of the day, it comes down to this: If insurance companies refuse to cooperate in health care reform, this nation will bypass insurers completely in favor of a government single-payer plan.

"The marvelous richness of human experience would lose something of rewarding joy if there were no limitations to overcome. The hilltop hour would not be half so wonderful if there were no dark valleys to traverse."

Helen Keller

CHAPTER 8

Important Factors in Health Care's Future

In her book, *What Makes Entrepreneurs Entrepreneurial?*, author Saras S. Sarasvathy describes two types of reasoning: effectual and causal. Effectual reasoning drives entrepreneurs by beginning, not with a specific goal but instead, with a means or skill that is capitalized and exploited into a business or undertaking. In focusing on a skill or way of doing business, the entrepreneur hopes that the outcome will be favorable in one or more ways. We can see Adam Smith's invisible hand at work here.

Causal reasoning is quite different and provides the power for professionalism. Causal reasoning begins with a visualized goal—such as health care for everyone. The reasoning is directed toward achieving the goal. Any number of means may be used to achieve the goal. Since we are exploring how to achieve health care for everyone, using causal reasoning, let's look at some factors that will impact how we achieve our goal.

Information Technology

You may get what you measure. But, possibly more important, you probably don't get what you don't measure or maybe don't know to measure. So, we require information technology for a successful health care system. We must ask the right questions and then probe for the correct answers.

Information technology has impacted health care more slowly than it has impacted business where the changes have been dramatic. However, electronic medical records are finally appearing in the clinical areas. Although electronic medical records are becoming popular, substantial costs and technical obstacles prevent saturation. About 25 percent of physicians' offices have installed electronic medical records. Some physicians have purchased electronic medical records systems, only to find they didn't work. To help sort things out, the federal government is moving towards approving and certifying selected electronic medical record systems so that physician providers can be sure of their utility, standardized use, and eventual interoperability.

The spread of information technology in health care is moving at a pace much like the development of computers--continuous but not uniform. Some medical people are very advanced in computerized information gathering and others haven't even started. But it is clear that information technology is and will be extremely important in the reform of health care.

Reforming health care has a basic premise: good health care can be expensive or less expensive or even reasonable in cost, but bad health care is *always* expensive. Creating and utilizing transparency in cost, performance, outcome, and complications becomes a powerful stimulus for all providers—doctors, hospitals, therapists and drug companies—to improve.

The Alegent Health Company describes the future this way.

"The next generation of health care will be the consumer's generation of health care, where patients are firmly in the center of any health care equation surrounded by transparency in quality, transparency in price, access solutions and innovative incentives. Here they will have the choice and control to engage in their health and make good health care decisions."

By using information technology to gather and then post statistics about costs, performance quality and outcomes, all players in the health care system stand to benefit.

The American College of Surgeons recently reported, "In response to educational programs, surgeon performance in clinical trials has measurably improved. Quality assurance audits have served both surveillance and educational roles."

Here are the three blue ribbon consensus recommendations that pertain to information technology:

- Increased transparency—information sharing—among systems and physician practices (Mayo doctor's group)
- Expanding the use of interoperable information technology (Commonwealth Foundation)
- Increasing transparency and reporting on all quality and costs. (Commonwealth Foundation)

Information technology will provide more than electronic medical records. When mature, it will include not only information on individual patients, their diseases, their treatments and their outcomes but also on physicians and hospitals together with their quality and performance data.

The next tier of development will be universal computer compatibility for providers. This long-awaited development will offer provider transparency, accountability and appropriately recognize those who excel.

Information technology, when mature, could:

- Decrease loss of medical records and misplaced tests
- Allow access to medical records from all parts of the United States
- Allow portability for long moves
- Define quality and best practices through the study of therapies and outcomes
- Save money by abandoning (or controlling) poor and ineffective treatments including pharmaceuticals
- Publicize best performers and best therapies
- Increase the inexorable pressure for providers to improve
- Define realistic and effective basic insurance packages that could be mandated
- Increase transparency and accountability throughout the health care system
- Facilitate managing the cost of a universal health care system
- Measure and publish prices including what procedures and services are covered in each episode of care
- Offer a standard connection that allows all health system information to be quickly and securely communicated and exchanged as necessary
- Manage and publish prices
- Measure and publish benchmarks defining quality care
- Create positive incentives to reward both those who offer and

those who purchase high quality, competitively priced health care.

While these advantages will take time to materialize, all are possible. In the meantime, it is a good idea to keep a copy of your own medical record so that you can carry it with you.

Demographics

Demographics is not exactly a means to help us get universal health coverage. Yet it is a driving force toward the goal. Americans are getting older. Predicting the future is difficult but we do know quite a lot about the future of the population of the United States. As discussed earlier, the massive Baby Boom generation, maturing at a known rate, will inundate the facilities that care for older people. Boomers will double the enrollment in Medicare. Facilities will need to expand to care for the growing of the aged population. Integrated, planned health systems may help with this problem.

In the same way, our children are urging us along, too. Children are our future. Yet approximately 9 million children do not have health care coverage. That represents a large drag on future U.S. health status.

Politics, Ideology and Philosophy

It is clear that in the political arena, Republicans and Democrats look at health care differently. The perception of socialism, real or imagined, always comes up in a discussion about universal health coverage. As noted before, socialism is defined as government owning and operating the means of production. This does not apply to mandated health care for all, when the providers are both privately and publicly owned. It is true that capitalism makes no claim to benefiting everyone equally. In health care, it is not possible to make valid delineations using economic class or ability to pay as criteria for eligibility.

One major advantage of the mandated, private, individually owned basic health care policy is that it appeals to conservatives because it's privately owned and nongovernmental; it appeals to liberals because everybody is adequately covered. Libertarians are not happy about the mandate but since health care must be provided, either the individual pays or the taxpayer pays.

Those who push for states' rights would have to acknowledge that the nationally administered Social Security and Medicare work better and more reliably than do state programs and Medicaid.

This brings to mind the words of Alexis de Tocqueville, who in

1832, said, "The business of the union is incomparably better conducted than that of any individual state."

U.S. Economics

The economics of health care and the economics of the United States are intimately related. The United States spends more than $2 trillion per year on health care. This amounts to about 16 percent of our gross domestic product.

Lowered federal taxes and the high costs of the Iraq war have decreased U.S. government disposable income. This has led to cuts in Medicaid health care services and Medicare reimbursements. The federal government faces sizeable deficits, current and future. Medicare's debt is an estimated $32 trillion, according to the Weiss Research Company's annual trustee report. This puts the government in a very poor position to accept additional financial burdens that health care reforms might impose. We don't seem to comprehend how new lower income tax rates, and thus lower tax revenues, are going to limit our opportunities to improve health care. In fact, the current financial trajectory for Medicare will either bankrupt the nation or require severe cuts in Medicare benefits.

Health care is going to remain an important segment in our national financial picture. Insurance companies and medical professionals are not going to willingly give up any of their income. What's reasonable is to demand better value for the money being spent.

Another factor is the effect of increased demand for limited commodities like oil as the world's population increases. This will increase commodity prices, which in turn puts pressure on personal budgets. Prices for all goods continue to rise because of scarce raw petroleum and shortage of operational refineries. It is time for us to demand the best value for our money in all areas, and that includes health care.

Health Care Economics

Health care is an important and integral part of the U.S. economy, just as it is a part of most nations' economies, worldwide.

So why do we regard health care as a burden?

Admittedly, the question is difficult to answer in the positive because, at first glance, health care can appear to be a drag on the U.S. economy. However, consider this:

- Health care is not easily exported in globalization (although we do see Americans going abroad for expensive treatments).
- Health care in 2006 amounted to 16 percent of our gross domestic product. This, in itself, is a generous contribution to the economy.

Then too, as the population ages and as life expectancy continues to increase, we will need healthy people, working longer, to contribute more to the U.S. economy. Excellent health care directly addresses this, by keeping people healthier so they can work, and by providing jobs for people to work at. These benefits contribute greatly to the economy.

As a side note, the United States' life expectancy in 2001 ranged from 66.6 years to 81.3 years, varying by area of the country. In general, the southern states had the lowest life expectancy, while the northern states fared considerably better.

But let's return to the question of whether or not health care is a burden on the economy. True, health care costs are rising, especially in the U.S. Let's look at the inflating influences on health care and see how we can improve as many as possible. To begin, let's identify what is fixed.

We cannot easily improve these causes of increased expense:
- The best health care in the United States is also the best health care in the world. The best medicine, the best physicians, the best research, and the best hospitals are costly.
- Continuous development of new drugs and new technologies is costly.
- A massively aging population will require more care. (But note, it was medicine that helped so many people stay alive.)

We *can* improve these causes of increased costs:
- Use the newest drugs and technologies. We need a method of adopting and dispersing appropriate new developments, but with discernment as to the value of each innovation that comes along.
- Unopposed pharmaceutical pricing is allowed. Our government, unlike Canada, Great Britain, and France to name a few, does not negotiate favorable pharmaceutical prices as does the rest of the developed world. This certainly is a concern. Yet, if all drug company profits were refunded, health care costs would fall by only about 1.2 percent.
- Competing hospitals and competing physician groups create

relentless modernization pressure. But who is paying for this? Hospitals typically operate on a slim profit margin of 5 percent or less.

- The current government and insurance payment schemes reward physicians for volume over quality or outcome.
- Nearly 50 percent of medical activity is judged to be either inappropriate or wasteful, according to a 2003 study published by the Institute of Medicine.
- Poor lifestyle choices also contribute, resulting in obesity, addictions to tobacco and alcohol, and a sedentary lifestyle. These correctable conditions cause expensive-to-treat illnesses.
- Our uninsured population now must have health emergencies cared for in emergency rooms, which is the most expensive way and not applicable for chronic diseases.
- A large reservoir of uninsured, untreated and unmanaged chronic medical conditions are destined to develop into expensive acute illnesses with long, costly treatment and poor recovery prospects.
- Large overhead costs in the private insurance sector now approach 25 percent. It is expensive to determine what is or is not covered with each claim, and to protect insurance companies against adverse selection.

Moral and Ethical Considerations

Starbucks CEO Howard Schultz declares that health care is a moral obligation. And, of course, the physicians' professional Codes of Ethics require the interest of the patient to be the doctor's first concern. Physicians are expected to adhere to these codes of ethics which specify the proper relation and conduct between themselves and their patients. Yet economic and other interests intrude and compete.

It is not easy to live 100 percent up to the code of ethics that requires the interest of the patient to be foremost because medicine is also a business.

There is a tension between market-driven and egalitarian medicine. Some maintain that the market can solve the problem of health care for all. It is not immediately clear how and as yet, it hasn't happened.

In 1983 the President's Commission for the Study of Ethical Problems in Medicine and Biomedical and Behavioral Research proposed: "Health care institutions....have a responsibility to ensure that there are appropriate procedures to enhance patients' confidence, to provide for the designation of surrogates, to guarantee that patients are adequately informed, to overcome the influence of dominant institutional biases, to provide review of decision-making, and to refer cases to the courts appropriately."

Research protocols can be very complicated and possibly involve impaired capacity research subjects. The ethics committees in hospitals and research institutions will find themselves facing many difficult problems.

The American Medical Association Principles of Medical Ethics were developed primarily for the benefit of the patient. These principles are standards of conduct and honorable physician behavior.

Wouldn't you like to know what you should expect from your physician? Well, read on. The AMA Principles of Medical Ethics states:

I   A physician shall be dedicated to providing competent medical care, with compassion and respect for human dignity and rights.

II   A physician shall uphold high standards of professionalism, be honest in all professional interactions, and strive to report physicians deficient in character or competence, or engaging in fraud or deception, to appropriate entities.

III   A physician shall respect the law and also recognize a responsibility to seek changes in those requirements which are contrary to the best interests of the patient.

IV   A physician shall respect the rights of patients, colleagues, and other health professionals, and shall safeguard patient confidences and privacy within the constraints of the law.

V   A physician shall continue to study, apply, and advance scientific knowledge, maintain a commitment to medical education, colleagues, and the public, obtain consultation, and use the talents of other health professionals when indicated.

VI   A physician shall, in the provision of appropriate patient care, except in emergencies, have the freedom to choose whom to serve, with whom to associate, and the environment in which to provide medical care.

VII   A physician shall recognize a responsibility to participate in activities contributing to the improvement of the community and the betterment of public health.

Professional Behaviors of Surveyed Physicians, Selected Measures	
Question	**Percentage of Respondents**
Honesty with patients In the last three years, have you told a patient's family member something about a medical issue that wasn't true?	less than 1% answered "Yes"
Improving access to care In the last three years, have you provided care, with no anticipation of reimbursement, in a setting serving poor and underserved patients?	less than 74% answered "Yes"
Improving quality of care In the last three years, have you participated in a formal medical error reduction initiative in your office, clinic, hospital, or other health care setting?	less than 53% answered "Yes"
Maintaining trust by managing conflicts of interest Scenario: You and your partners have invested in a local imaging facility near your suburban practice. When referring patients for imaging studies, would you: 1. Refer your patients to this facility? 2. Refer your patients to this facility and inform patients of your investment? 3. Refer patients to another facility?	less than 24% selected answer 1
Fulfilling professional responsibilities, including self-regulation In the last three years, have you had direct personal knowledge of a physician who was impaired or incompetent in your hospital, group, or practice? If yes, how often did you report that physician to a hospital, clinic, professional society, or other relevant authority? 1. Always 2. Usually 3. Sometimes 4. Never)	45% selected answer 2, 3, or 4, indicating that they had not reported at least once

Source: Adapted from E. G. Campbell, S. Regan, R. L. Gruen et al., "Professionalism in Medicine: Results of a National Survey of Physicians," *Annals of Internal Medicine*, Dec. 4, 2007 147(11):795-802.
Chart from The Commonwealth Fund: www.commonwealthfund.org/publications; Summary Writer: Deborah Lorber.

VIII A physician shall, while caring for a patient, regard responsibility to the patient as paramount.

IX   A physician shall support access to medical care for all people.

Leading the Charge

Leadership for change is in scarce supply when the status quo is rewarding. One would think that physicians would be the natural leaders in obtaining health care for everyone, but it hasn't happened yet.

Dr. James T. Dove, president of the American College of Cardiology observed, "Some physicians would prefer to be in the back of the room and hope they're not called on. That is not an option. Physicians must step up to the plate to preserve what is good about the health care system while correcting its deficiencies."

On the other hand, business leaders are also adversely affected by our present health care policies and structures. Those with power and financial impetus are taking action. The New American Foundation is working with McKinsey & Co. to recruit CEOs willing to help publicize the cost, quality and coverage crises in our present system. Business leaders and their companies want very much to get out of the present health insurance business, so it's clearly to their advantage to lead in health care reform.

To cut costs, some companies are installing heath care centers offering primary care in the workplace, and hiring staff physicians to attend as often as the demand requires.

Businesses are using research to find the best specialists and the best prices for referrals, and they are offered those practitioners 20% higher fees to take the corporate referrals.

Tort Reform

The United States has no organized way to help people who receive poor results in the health care system. The only recourse is through the legal system, already sorely overburdened. Some patients do sue their physicians, although juries often decide in favor of the physician. But when judgment is in favor of the patient, the monetary awards can be unpredictably high.

In the United States, the issue of tort reform is a very hot topic for physician organizations because malpractice insurance, which protects the physician, has become prohibitively expensive.

Since the benefits of the system are so sparsely distributed and yet so expensive, it is reasonable to propose changes.

- The most idealistic approach is to increase patient safety and reduce medical errors, in order to reduce or eliminate remedial action of any kind. The U.S. medical community is devoting considerable effort to this goal.
- Another approach consistently favored by physicians is to lobby for state-legislated caps, or upper limits, on non-economic medical damage claims.
- A third approach is used in Sweden. Any patient who sustains a significant medical injury or a costly poor result receives a commensurate monetary benefit from a government board which is charged with evaluating the complaint and making such awards. If a physician is suspected of malpractice, an entirely different government board reviews the matter and, if warranted, can impose a penalty on the physician. The major point here is the separation of patient financial compensation from professional or staff discipline.

Things alter for the worse spontaneously,

if they be not altered for the better designedly.

Francis Bacon

CHAPTER 9

Effects on the Players

It is clear that not all providers, insurers, employers and new makers will cheerfully accept reforms. Some will understand and embrace reforms that benefit patient welfare. Others will, in varying degrees, object, obstruct, and otherwise cling to the status quo. But change, like it or not, is on its way.

Let's consider how our proposed, private, mandated, insured universal coverage (PMIUC) might possibly affect each of the players.

PMIUC Effect on Player 1: The American public, for whom health is the goal.

First and foremost, with PMIUC, everyone will be covered. This will appeal to those who are currently uninsured and poorly insured. Political liberals will like it as a necessary and laudable benefit for people. For conservatives, it's private, not operated by the government.

In what ways will PMIUC impact the American public?

- The goal of universal coverage will be achieved.
- 47 million new, basic health care policies will be issued
- No worries and no rejections for pre-existing conditions.
- All patients have basic insurance to pay for care.
- No tiers of care in the basic policy.
- Each person may take his coverage with him from job to job.
- Each person may select his own insurance company instead of the company chosen by employer.
- Individual ownership of health care policies creates a new, valid relationship with insurance companies. Each American, not the employer, is now the insurer's proper and appropriate customer.
- Insurance companies' competition on cost and quality will

benefit the public.
- People are free to change insurance companies, for any reason.
- Americans will have more information, through information technology, about the types and quality of providers.
- With all Americans participating in the basic health care package, each individual's expense will be minimized.
- Employees are no longer held captive by companies because of fear of losing health insurance.
- Medical bankruptcy will be virtually eliminated.
- Coverage is to be documented with the annual income tax. Those not covered would be automatically enrolled in a default private plan and either billed or subsidized accordingly.

In the short term, possible delays may ensue, with increased waiting times for health care, fueled by previously untreated needs. But enrolling people in the month of their birthday may help smooth this out.

Under PMIUC—or any health care reform or no change at all—you and I, also known as the public, need to become active, informed health care consumers. Such intelligent involvement will help us shoulder our share of the goal of achieving good health. All, or most of us, need to participate in the burden and utilize best health practices. That's the only way it will work.

According to Christopher Murray of the Harvard School of Public Health, money plays a role in health, but our habits have the most impact overall. What are the most important habits for us to examine? The key risk factors are: tobacco, alcohol consumption, obesity, elevated blood pressure, high cholesterol, low fruit and vegetable intake, and physical inactivity. Americans are diverse, ethnically, and it's interesting to note that Asians, strengthened by their societal habits, live the longest.

What else do we need to look at? Our ability to read and understand health instructions is another big factor. Five out of six adults are confused by health instructions. The 2003 National Assessment of Adult Literacy concluded that adults over 65 are less literate—that is, cannot read and write well—than younger people. Blacks, Hispanics and American Indians have lower average health literacy than whites and Asian adults.

The truth is that very few people understand how the health care system works and how it should work. It is complex and often poorly described by those within the system. Some people are paralyzed by fear and avoid health care as long as possible. Some people are poor and can't afford to participate in their own care. Ideally, Americans should understand their health care system. It is by understanding the care that we receive that we can act responsibly on behalf of our own best health interests to get the recommended examinations and treatments.

The Mayo Clinic National Symposium on Health Care Reform recommended, for the American public: Reward patients who select high-quality providers.

But how do we find and select those high quality providers? What information could we tap to educate ourselves? Some of this information is out there already. In Chapter 14, you'll find a list of websites that provide information for you.

In addition, to help educate the public, the American College of Surgeons launched a patient education program in 2004. Its goal was to help patients and their families understand the specific surgical operation being planned and to have the knowledge and skills necessary to fully participate in the hospital care and continued care post discharge. Patient education materials were developed for mammograms, flu shots, smoke-free living, diet, weight control, exercise, and the need for regular screenings like colonoscopy and pap smears.

PMIUC Effect on Player 2: Health care providers which includes doctors, hospitals, health care professionals, and drug companies.

Doctors and nurses understandably worry about their positions when dramatic sweeping changes are about to take place. As the medically trained and knowledgeable people in the equation, these players need to be educated and reassured. Regardless of changes in process and payment, the need for doctors and nurses won't change.

The physicians' best stance is to take a leadership role, influencing changes to benefit patients and participating in decisions. As the plan is implemented, doctors will improve their practices and information technology will improve their therapeutic algorithms and the quality of care. The most successful practitioners will increase their market share. As time passes, the resistant, status quo physicians will come along or else fade away.

Admittedly, private, mandated, insured universal coverage will challenge the doctor. Being evaluated on performance can be a shock.

After all, a physician never stops learning—first at medical school and residency, then through continuing education classes, experience in a practice and reading current journals. The doctor applies this knowledge, to the best of his or her ability, to the problems of the patient. But medicine is complex. The doctor must remember a long list of best practices as he/she determines the best treatment for each patient whose circumstances are unique. Then the physician must depend on the patient to comply fully with the prescribed treatment.

So, yes, it comes as a rude surprise for some physicians to learn, through information technology, that their performance and outcome ranking is not up to the highest standard. But most physicians want to and, in fact, think they are doing a good job.

Others may disagree about that evaluation. In fact, a 2006 study suggests that physicians have a limited ability to accurately self assess. Medical personnel and systems are under constant pressure to improve. Upon learning that a better way is available, and under observation and pressure, most physicians will improve their performance.

The PMIUC plan will affect health care providers in these ways:

- Improved performance. As information technology precisely identifies best practices, providers will improve and cost-effective performance will accelerate.
- Leadership. Physicians will have opportunities to lead in the development of new practice parameters and standards for the welfare of patients. Physicians who lead will be rewarded.
- Pay for performance. Doctors and hospitals will be paid for outcomes rather than volume.
- Collaboration. Physicians and other providers will work together, in order to produce the best possible outcomes.

Group practice

Group practice originated at the Mayo Clinic in the early twentieth century. This form of practice has become widespread but is by no means universal.

Dr. Caldwell Esselstyn defines group practice as "the formation of a team of physicians with separate skills but common philosophies, who pool their skills for the benefit of the patient and who organize themselves so as to have a mutual responsibility both to their patients and to each other."

Group practice offers significant advantages for both patient and physician. The group can afford to hire excellent business managers and staff, allowing physicians to concentrate on medical care, to the advantage of the patient.

One group already successfully offering care is Kaiser Permanente. The largest integrated health care system in America, covering 8 million people in nine states, Kaiser spends 93 percent of its premium income on patient benefits.

Other advantages arise from a not-for-profit, integrated health care group practice such as the Mayo Clinic and Kaiser Permanente. In a group practice, physicians are…

- In charge of their practice environment
- In control of their hospitals
- Rewarded, because of excellence, by donations for better equipment, research and teaching
- Respected because of their devotion to patient benefit
- Well-paid
- Supported by hiring the best lawyers, accountants, investors and advisers.

The disadvantages, for some physicians, include a loss of independence and the need to be a team player.

Will the PMIUC plan improve disease management?

Physicians already have demonstrated conclusively that concentrating on the care of a single disease produces superior results. Specialty hospitals, owned and operated by physicians, give excellent results.

The U. S. General Accounting Office studied specialty hospitals in 2003, concluding that most fell into five types—cardiac, orthopedic, surgical, women's, and other. The other-specialty category included a variety of areas, such as eye and ear, nose, and throat procedures.

Third party payers are attracted to the good results and the reasonable cost of good care achieved in specialty hospitals. It is clear that poor medicine, with complications and poor results, is always expensive.

Will hospitals change?

Hospitals need an additional layer of care corresponding to the unfulfilled promise of a skilled nursing facility.

Both hospitals and physicians are providers. As such, they have many interests in common but often they are at odds with one another.

How do hospitals differ from doctors?

- Hospitals would like all their beds full. Physicians would like to keep some beds open so they can easily get patients in when necessary.
- General hospitals' facilities and routines do not work to the satisfaction of all specialist physicians. As a result, many specialists have formed their own small hospital or outpatient center where they control and choose the equipment and set up the procedures. This results in a financial strain on the general hospital that lost their specialist. But it frequently results in a high-quality center that offers excellent care for a specific illness.

Similarly, large, disease-specific hospitals such as M. D. Anderson Cancer Center or a bone and joint hospital or a women's hospital do attract appropriate specialists who, in turn, are the reason for the success of the hospital.

The Mayo Clinic and the Cleveland Clinic are other examples of very successful practice arrangements where the physicians are in charge and the hospital conforms to their needs.

So, both kinds of hospitals—general and specialty—are needed. Hospital administration and hospital personnel, by themselves, can ruin a hospital but, by themselves, cannot ensure the success of the hospital.

How will payment to providers change?

- The way to improve quality of care is to pay providers for outcomes rather than for units of work, as we do now with fee-for-service. Paying for good outcomes has several benefits:
- Patients receive good value, improvement, bundled services, and focused improvement in the health care segments that now cost the most.
- Financial incentives will be completely transparent for all players.
- Public reporting will increase value.
- The metrics, thanks to information technology, will be current and valid.
- Pharmaceutical companies will be under pressure to change their financial reward structure away from high-priced blockbuster drugs towards more generics and services with more modest financial returns.
- Unpaid bills and collection problems should be diminished or eliminated.

Will physicians behave differently?

Dr. Aaron Lazare of the University of the Massachusetts Medical School describes the art of apology as a profound healing behavior. He and others like him encourage physicians to disclose medical errors, acknowledge responsibility, offer explanations, express remorse, and offer an apology. Remarkable benefits accrue from humility.

PMIUC Effect on Player 3: Insurance companies, which manage risk

Under the PMIUC plan, insurance companies remain a player in the health care game. This is unlike a single-payer system, with the federal government assuming the role of insurer. So the Plan will impact insurance companies in the following ways:

- 47 million new insurance policies will be written
- Actuarial costs will drop, because all applicants are accepted. No pre-existing conditions or risk selection apply to the basic policy.
- Underwriting supplemental health care policies will continue as demand dictates.
- The national risk pool will enable actuarial calculations that are profitable, even with adverse selection
- Overhead will be reduced due to the uniform, universal basic policy
- Insurance companies will compete based on cost and quality of service
- Insurers will be encouraged to invest in disease prevention and long-term preventive care
- Excellent customer service will be rewarded with increased market share, as consumers flock to those companies
- Regulations that require 85 percent pay out will help with risk avoidance

Our blue-ribbon consensus groups made the following recommendations concerning system reform. These recommendations, under the proposed PMIUC plan, would impact the insurance industry:

- Reimburse health care based on results. (Mayo Clinic symposium)
- Reward performance for quality and efficiency. (Commonwealth Fund)
- Implement major quality and safety improvements. (Commonwealth Fund)

- Fundamentally restructure the way end of life services are financed and provided. (Citizen's group)
- Promote efforts to improve quality of care and efficiency. (Citizen's group)
- Define core benefits and services for all Americans. (Citizen's group)

Incentive Reinsurance

Reinsurance is necessary in some form. The form that seems to make the most sense mathematically is a national risk pool, as explored in Chapter 7. This allows each insurance company to handle a normal distribution of benefit payments. If an insurers encounters an extra heavy expense load, the national risk pool can restructure and ameliorate the load.

As we've seen, the top 20 percent of insured people account for more than 82 percent of the money spent on health care. Only if the unexpected occurs—if one of the lower deciles uses as much benefit money as the highest deciles use—would the total of benefits rise above income.

PMIUC Effect on Player 4: Government

Under the Plan, the federal government is charged with helping those who cannot help themselves. Here are some specific effects of the plan on government.

Congress will mandate a basic health care policy, based on the advice of a blue ribbon committee and the appropriate recommendation of new makers. Congress will authorize the blue ribbon committee to update the basic health care policy as needed

Congress will determine the method of payment and means tested financial support required for lower income citizens to participate in national health care.

All current monitoring and regulating agencies will continue their current duties and will also oversee the new national health care system.

The Plan will help the government by not expecting it to bear the tremendous expense of a completely government financed health care system for everyone.

Why would all this be in line with public and private expectations? Well, let's look at the blue ribbon consensus recommendations for the federal government:

- Establish public policy that all Americans have affordable health care. (Citizens group)
- Expand health insurance coverage to all residents. (Commonwealth Fund)
- Encourage public–private collaboration to achieve simplification, more effective change. (Commonwealth Fund)
- Fundamentally restructure the way end of life services are financed and provided. (Citizens group)

It's important to consider the national debt. For a variety of reasons, the United States government is in a substantial debt condition. It seems very unlikely that the federal government would willingly add an enormous financial liability such as a health care system that is completely financed by federal funding.

Various institutions of the government are funding current research that will help develop quality measurements for health care, which are necessary. The job is not yet over. It will take a lot of work and wisdom to develop adequate and complete metrics.

PMIUC Effect on Player 5: New Makers

New Makers is my term for those who develop the innovations and breakthrough ideas that improve health care and medical practice. These new practices and ideas continue to come from all areas: medicine, pharmaceuticals, technology, and government. It is to all players' advantage to see that the best contributions of New Makers are incorporated into health care, promptly and realistically.

Our blue ribbon consensus groups said this about new developments.

- Encourage public-private collaboration to achieve simplification, more effective change. (Commonwealth Fund)
- Foster innovative integrated community health networks. (Citizens group)

Right now, many breakthroughs occur but few physicians and hospitals are aware of them or are motivated enough to adopt them. Under the proposed plan, New Makers will be reviewed in real time by a knowledgeable panel of experts, and those selected as best practices will be widely publicized to all players, including the public.

The early stages of this are already happening. In 2006, the National Institutes of Health (NIH) awarded grants to 12 educational institutions nationwide to develop a consortium. According to NIH Director Elias Zerhouni, M.D., in Washington, D.C., "The development of this consortium represents the first systematic change in our approach to clinical research in 50 years." Zerhouni also said, "Working together, these sites will serve as discovery engines that will improve medical care by applying new scientific advances to real world practice. We expect to see new approaches reach underserved populations, local community organizations and health care providers to ensure that medical advances are reaching the people who need them."

One of the institutions in the new consortium is Duke University Medical Center, which formed the Duke Translational Medicine Institute (DTMI). DTMI includes a special arm that is focused on streamlining the process of guiding new scientific discoveries through the early phases of development into technologies that can be applied to human health.

The United States is respected for advancements and breakthroughs in both the tangible and intangible areas of medical science. The unfortunate flip side is that the new technologies, in the past, did not become widely available to ordinary patients.

In the intangible realm, the New Makers include new ideas, new considerations, new quality measures, new laws and new considerations of every kind. While some are intangible, these revolutionary approaches to medicine and health can have every bit as much impact as new medical devices, drugs, and procedures.

For successful health care reform, we must develop a mechanism to evaluate the innovations, then implement those offering best value. I had envisioned a central agency or several competing agencies to evaluate and recommend the adoption and integration of the New Makers' efforts. The new consortium being developed by the NIH has exciting potential to serve this very purpose.

PMIUC Effect on Player 6: Employers

The proposed plan allows the employer the option of withdrawing from health care, unless a business reason exists to offer it. An employer is free to pay for employees' basic policy, of course, or to pay an appropriate wage to attract employees.

Ownership of the policy is in the individual's hands, in either case. Businesses are no longer pressured to offer employee health care insurance.

An element of fairness enters the workplace. In today's world, a wife or husband, employed by a company offering no benefits, can use benefits from the spouse's employer. Thus, the employer who offers benefits is actually paying for those employers who *don't* offer coverage. In the same way, businesses that do not provide insurance will no longer depend upon taxpayers to pay for their employees' health care through state and federal health programs.

Finally, under the PMIUC plan, health care's drag on U.S. businesses' global competitiveness is removed.

Will players universally applaud the new PMIUC plan?

It is unlikely.

As Garrett Hardin, author of *The Tragedy of the Commons* mentioned previously, pointed out, "It is one of the peculiarities of the warfare between reform and the status quo that it is thoughtlessly governed by a double standard. Whenever a reform measure is proposed it is often defeated when its opponents triumphantly discover a flaw in it."

Kingsley Davis, sociologist and U.N. Population Commission member, agreed. "Worshippers of the status quo sometimes imply that no reform is possible without unanimous agreement, an implication contrary to historical fact. As nearly as I can make out," Davis commented, "automatic rejection of proposed reforms is based on one of two unconscious assumptions:

- the status quo is perfect; or,
- the choice we face is between reform and no action; if the proposed reform is imperfect, we presumably should take no action at all, while we wait for a perfect proposal "

There is no perfect proposal that all can agree upon for universal health care. We must rely upon mutual coercion, mutually agreed upon. It will be the job of reform to help society define the mutual coercion that we can mutually agree on that will serve society best.

America's most urgent and visible need is universal coverage. Offerings that are less than universal will ultimately, if not immediately, be problematic.

This process reform will address the how of system reform. It will address how health care might be delivered and how it might be paid for.

CHAPTER 10

What about a Single Payer System?

The single-payer system would be better than the incremental system we have. Its major advantage? A single-payer system is a bumper sticker in its simplicity: the government pays for health care.

This brings us back to the concept of the basic health care package. You will recall that we explored the concept when discussing the Plan, the mandated individually owned policy. The basic package was referred to by our blue-ribbon consensus panels.

C4 Define core benefits and services for all Americans.

F1 Expand health insurance coverage to all residents.

M5 Define essential health care services.

Whatever the basic package is determined to be, it could and should be the same concept whether it is generated for a single-payer system, the private, mandated insured universal coverage we call the Plan, or any other system. So the extent of coverage—the basic package—need not be the factor that decides the system.

The closest thing we have to actual details of a basic health care package comes from Rep. Dennis Kucinich's Congressional 2007 proposal in HR 676:

- All Americans citizens receive benefits
- Primary and preventive care
- Inpatient and outpatient services
- Emergency room services
- Prescription drugs

- Long-term care
- Mental health services
- Dental coverage
- Hearing and vision coverage
- Substance abuse treatments
- Chiropractic services
- No deductibles, no co-pays, no out-of-pocket expenses
- Portable; see any doctor in United States.

It's this basic package, devised by Rep. Kucinich for a single-payer system, that we introduced earlier as one possibility for the PMIUC plan's system utilizing mandated, individually owned basic policies.

Single Payer Payment

Here's how Rep. Kucinich would pay for a federally controlled single payer system of health care:

First, he would use already-earmarked U.S. government funds which currently pay for 47 percent of health care. This includes Medicare, Medicaid, Veterans Administration, elected officials, military personnel, and a few others.

Next he would tax all taxpayers 2 percent of payroll taxes and raise the tax on the upper 5 percent of taxpayers.

Finally, Rep. Kucinich would tax employers approximately 7 percent of payroll. Currently employers pay about 25 percent of payroll so this represents a considerable savings.

Advantages of Single Payer

There are advantages to a single payer system. Here's a list of those that come to mind:

- It is not cumbersome; it is simple
- Everyone is covered. In this sense, it is in the public interest compared to our current system, as we've already discussed, where insurance companies behave in ways that are not in the public interest. However, this is not better than the privately mandated Plan we've proposed which also covers everyone as well as forcing public service behavior from insurers.
- It's better than what we have.
- Initial enthusiasm is high.

- Many physicians support this plan.
- Adverse selection is eliminated.
- Cost shifting is eliminated.
- Employers are out.
- Asymmetrical information is eliminated.
- Other countries are using the system and can serve as models.
- The Plan would be managed by the National Board of Universal Quality and Access.
- Some physicians will be salaried.
- Regional, state and national boards will determine hospital budgets.
- Hospitals can reduce their bureaucracy that formerly dealt with the insurance industry.
- Insurance companies could no longer profit by cherry picking the best cases or increasing premiums on expensive patients.

Disadvantages of Single Payer

Having said that, the disadvantages of a single payer health care system are enormous. Please consider them carefully. Several of them are deal-breakers, in and of themselves.

First, stakeholders such as the insurance industry with its big budgets and enormous lobbies will go all out to fight such a radical change. In 2003, two of Washington, D.C., largest health insurance trade groups—The American Association of Health Plans (AAHP) and Health Insurance Association of America (HIAA)—merged. Their estimated combined budget at that time was $40 million, according to the organization called Physicians for A National Health Program. The two insurance lobbyist organizations were said to "own a controlling interest in Congress." We have no reason to believe that control has lessened in 2008.

This radical change to a single payer system will take a long time to complete, given the opposition. of powerful insurance lobbies. Remember the ill-fated Clinton health care reform plan of the 1990s. The same kind of vicious fight will ensue when a single payer system is proposed. This could drag out for more years while more and more Americans suffer from lack of affordable health care.

The second major disadvantage is that funding money must come from reluctant legislators in Congress every year. This annual fight for funding is Canada's biggest problem with their national health plan. In this country, remember what the federal legislatures have done to Medicaid.

And of course, universal health care under a single payer set-up is a very large entitlement. It's a given that that will be vulnerable every year at appropriation time to legislators who are opposed to entitlements.

A third issue is the abrupt entry of 47 million uninsured people into U.S. medical waiting lines. At the present time, the U.S. is often complimented for shorter waiting lines than Canada, but a single payer system could change that.

One of the most troubling issues is that a single payer system is monolithic, similar to a dictatorship, without competition on cost and quality. This allows a single payer system to be infinitely more susceptible to manipulation than a mandated system like the Plan, in which all companies compete on cost and quality to acquire and maintain market share. Single payer has no incentives for private competition, which is the mechanism for controlling costs.

Another issue is that the government chooses the insurance package. Some advocates of single payer, like John P. German, past president of Physicians for a National Health Program, believe this is better than mandated individually owned basic health care policies like the Plan. Dr. German says government control is better than what he calls "market-based insurance schemes." However, his description does not fit privately owned mandated policies owned by all, because the same basic policy must be offered by all insurance companies and they must accept all applicants, and all must apply. This does not fit the description of "market-based."

It's interesting that Dr. German, in his article, "Leading Dems Miss the Boat on Health Care," uses, as an example of political compromise and its hazards, the Medicare Prescription Drug, Improvement and Modernization Act of 2003 (MMA). Instead, analysis shows that it is actually a hazard of the government being manipulated by lobbyists of the drug and insurance industries. It is an illustration of how vulnerable to manipulation would be the single payer system's standard policy-- much more so than in a mandated system in which all companies compete on cost and quality to acquire and maintain market share..

In Congressman Dennis. Kucinich's plan, no cost controls are visible right now.. This is of concern because of the federal government's track record. Remember, Medicare is up two and one-half percent per year every year, more than the gross domestic product. Medicare's efforts to cut costs have resulted in bizarre distortions, such as cardiac surgeons being paid less for coronary bypass than ophthalmologists are paid for cataract surgery. The ultimate cost control can only come when information technology matures. Likely that will take a long time. In the meantime the government deficit is enormous.

Finally, a shift to a single payer federal health care system could result in the loss of jobs for thousands of health care insurance company employees. While it is unlikely that insurance companies as a whole will go out of business, given the many other products that most sell, the employees could suffer, with potential layoffs substantial enough to trigger a recession.

7-09-2007 @ 5:57PM

happyInCanada said...

I feel so terribly sorry for all you Americans....

I'm happy here in Canada.

Two knee operations, my dad had triple bypass

and a family member on

permanent mental healthcare.

All free.

Good luck to you

http://www.gadling.com/2007/07/05/what-countries-have-universal-health-care/

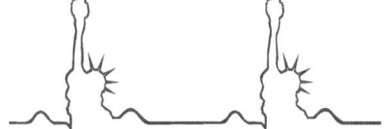

CHAPTER 11

Health Care in Other Countries

Australia	**Germany**
Belgium	**Italy**
Canada	**Netherlands**
England	**Sweden**
France	**Switzerland**

As you can see from the above list, we will look at a small sampling of countries and their health systems. This is not meant to be a comprehensive comparison. Rather it's an overview of several countries with which you may be familiar. These pages include statistics from varying timepoints, so please keep in mind this caveat as you read.

So let's begin by going to the next page and looking at our largest "neighbor" down under: Australia.

Australia

Population: 19.6 million
Life expectancy: 79 years
Infant mortality: 4 per 1000 births
Health care: About 8.5 percent of GDP

Australia's government-sponsored Medicare Australia is a single-payer system that covers all Australians and is supported by general tax revenue. It pays the entire cost of treatment in a public hospital. Outpatient charges are set by physicians but the government pays according to a fee schedule. About 85 percent of family physician and outpatient services are directly billed to Medicare Australia. About 75 percent of hospital services are billed directly to Medicare.

Private hospitals (about 3 percent) have about 30 percent of hospital beds. Public hospitals (93 percent) are owned by the federal or state governments and are the main location of emergency rooms. They are reimbursed with Diagnostic Related Group (DRG)-type payments.

The government pharmaceutical benefit has a co-payment for prescriptions.[29]

About 50 percent of the population has private, individually owned health care insurance policies. Private insurance companies receive subsidies from the federal government.

Health care costs are paid with about 47 percent from the federal government and 23 percent from state government. The remainder is paid by private insurance and private pay.

Australia offers good medical care in both public and private sectors.

Problems with Australian health care include escalating costs, waiting lines, malpractice insurance, hospital accidents, aging populations and access in rural areas.

Belgium

Population: 10,379,067
Life expectancy: 79 years
Infant mortality: 4 per 1000 births
Health care: About 10% of GDP

Belgium combines government and private insurance into what is considered one of Europe's best health care systems.

The government insurance is compulsory and part of the Social Security system (Health Insurance Funds). Employees' and employers' contributions are deducted automatically from salaries by the National Office for Social Security.

With government insurance, a patient can choose any doctor, clinic or hospital without referral.

Physicians who charge by a government schedule receive cash payment. This is followed by refunds of about 75% from government insurance using stickers attached to the bill. A standardized credit card is issued for payment at pharmacies and hospitals. Pharmacies are very accessible and rotate 24 hour coverage.

Private insurance companies offer coverage similar to the government insurance and to supplement government insurance.

Physicians may work for both government insurance and private insurance, but most work for one or the other.

Dentists are mostly in private practice although some accept payment from state insurance.

Belgium is the healthiest country in the world and has the second-highest physician-patient ratio.

Note: For a fascinating look at the similarities and differences between U.S. and Belgium health care systems, featuring the Belgium minister of health and learned American health care experts, held at the Brookings Institute in February 2007, go to:

http://www.brookings.edu/~/media/Files/events/2007/0205health%20care/2007 02_Belgian.pdf

Canada

Population:	**33 million**
Life expectancy:	**80 years**
Infant mortality:	**5 per 1000 births**
Health care:	**About 9.9% of GDP**
Doctor-patient ratio:	**2.29 doctors per 1000 inhabitants**

Canadians have the highest life expectancy in the developed world.

The Canadian health system is actually ten different systems, each run by a province and coverage for three territories. It is publicly funded and covers all necessary health services, including hospitals and physicians and surgical dentists. The provinces contribute 85 percent and the central government contributes about 15 percent. While several provinces collect premiums as well, the Canada Health Act prevents denial of service for inability to pay premiums.

Waiting periods for new residents cannot exceed three months.

The Canada Health Act is comprised of five main principles:

1. Public administration: All administration of provincial health insurance must be carried out by a public authority on a non-profit basis. They are accountable to the province or territory, and their records and accounts are subject to audits.

2. Comprehensiveness: All necessary health services, including hospitals, physicians and surgical dentists, must be insured.

3. Universality: All insured residents are entitled to the same level of health care.

4. Portability: A resident that moves to a different province or territory is still entitled to coverage from their home province during a minimum waiting period. This also applies to residents which leave the country.

5. Accessibility: All insured persons have reasonable access to health care facilities. In addition, all physicians, hospitals, etc, must be provided reasonable compensation for the services they provide.

Canadians receive a health card that provides coverage in a particular province. It is portable and can be used in other provinces for a period of time. After obtaining a health card, a Canadian registers with a primary care physician. For routine visits, the health card is sufficient.

Canada has about 30,000 primary care doctors in Canada and 28,000 specialist doctors. The physicians submit their bills to the province and are paid according to a fee schedule for the work performed.

Every covered person receives the same level of care. There is no upgrade insurance for the services covered by the Canadian government.

Waiting lines for care have been an issue, but that is a function of the annual appropriation level. A higher appropriation could shorten the waiting lines.

Private insurance is available to cover other services that the Canadian health system does not cover, such as regular dental care, optometrists and prescription medicines.

Private clinics with specialized services are available but federal law forbids offering services that are covered by the Canada Health Act. Regardless, many do offer such services without the wait. Private insurance may pay about 80 percent of those costs.

Canadians do have some problems:

- Federal interference in health care is not welcome.
- Private health care is considered a second-tier. While it shortens waiting lines, it does create an unwelcome two-tier health care system.
- There is a shortage of nurses and physicians.

England

Population   **50,500,000**
Life expectancy: **78 years**
Infant mortality: **5 per 1000 births**
Health care: **7.7% of GDP**

England has a tax supported, universal health care single payer system called the National Health Service (NHS). Created in 1948 and funded by general taxation, the NHS is run by the Department of Health with a government minister, the Secretary of State for Health, who is responsible for planning, regulation and inspection of the service. The NHS includes 28 regional health authorities and 300 primary care trusts that decide on the services to be provided locally.

The care is divided into primary (outpatient) delivered by 300 primary care trusts, and secondary (hospital care) delivered by a multitude of secondary trusts. Primary gets about 75 percent of the budget. Secondary trusts are: Acute Trusts for short-term care, Care Trusts for both health and social care, Mental Health Trusts, and Ambulance Trusts.

For various reasons, there is also demand for private insurance. This is usually purchased to cut waiting times or to improve service. While the UK has more than 300 private hospitals, Britons with a serious illness head for public services because that's where the best specialists are located.

Since 2002, the British government has created a very extensive information technology program called NHS Connecting for Health. Britain's is now the largest health care information technology system in the world.

NHS connects physicians' offices, hospitals, community and mental health organizations to a national information service. Each and every patient is issued an identification number to preserve anonymity. Clinical data can be transferred to and from all NHS organizations. A "spine" has information about every person in England. The services include booking, procurement, audit and cost control. Patients can access a summary of their health care.

There are, of course, technical problems. For the system to work, it needs to be usable and available from anywhere. Confidentiality and accessibility are under discussion. Three more years will probably see considerable improvement.[64]

A recent study (*Journal of the American Medical Association*, May 3, 2006, p. 2037)by English authors determined that U.S. residents are much less healthy than their English counterparts and that these differences exist throughout all socioeconomic groups. The study examined self-reported rates of several chronic diseases and then adjusted for age and health behavior risk factors as well as education and income.

France

Population:	**60,800,000**
Life expectancy:	**79 years** *(French women enjoy the second highest life expectancy rate of all women in the world)*
Infant mortality:	**5 per 1000 births**
Health care:	**About 10.1% of GDP**
Doctor-patient ratio:	**3 doctors per 1000 inhabitants**

France's complex health care system has evolved over one hundred years.

In June 2000, the World Health Organization declared France had the best health care system in the world. All legal residents—more than 96 percent of the population—have free or reimbursed health care. The French can choose among providers without regard to income.

The Social Security System, mostly for public health care, is supported by a 20 percent tax on workers, tax on tobacco and alcohol and income tax on all income sources. The legislature passes a budget each year based on revenues and goals to be achieved.

Low income people have free care. Eighty percent of the French have private insurance as well, to supplement providers' payments.

The system is not free at the point of service. When seeing a primary or specialist physician, one pays the bill to be later reimbursed approximately 70 percent. Ninety-seven percent of all French physicians adhere to a convention of fees. In order to be fully 100 percent reimbursed, one must be seen by a member of a group practice or have supplementary insurance.

France's system offers three kinds of hospitals:

- Public hospitals funded by public endowments.
- Not-for-profit hospitals, about 15 percent that were originally church-affiliated.
- Private clinics required to share medical files with public hospitals.

Other important agencies are:

- The National Institute of Health which supervises health care services.
- The French Agency of Health Safety of Health Products.
- French Institute of Blood.
- French Institute of Transplants.

Germany

Population:	82,500,000
Life expectancy:	79 years
Infant mortality:	5 per 1000 births
Health care:	About 10.5% of GDP
Doctor-patient ratio:	3.58 doctors per 1000 inhabitants

Germany has a mandated insurance health care system. State health insurance covers about 90 percent of the population that is below an annually determined income level. Premiums are collected from employees and employers. The remaining is covered by private, for-profit insurance policies or remains uninsured. If your coverage is private, you must have private long-term care insurance as well. If you are covered by statutory health insurance, you automatically have long-term care insurance.

Physician visits and prescriptions require co-payments.

The government decides on national health care policies.

Health care facilities in Germany are open to everyone. The system is not run by the federal government but rather by associations of payers and providers to one of which a German citizen must belong. The associations annually negotiate budgets and coverage with insurance companies—both public and private—and providers.

The German public is divided in its approval of the system. Problems include waiting lists for surgery, increasing costs and an aging population.

Italy

Population: **53,133,509**
Life expectancy: **80 years**
Infant mortality: **5 per 1000 births**
Health care: **About 6.5 % of GDP**
Doctor-patient ratio: 3 doctors per 5900 residents

Italy's national health system is known as the SSN, which stands for Servizio Sanitario Nationale. The SSN provides free or low-cost health care to all European Union citizens. It covers hospitalization (medicine and surgery and tests), family doctors, specialists, most drugs and medicines and dentistry.

Everyone is required to have some form of health insurance according to various rules. The employed get their health insurance through the employer and receive a health number and a health card entitling them to free visits with the family doctor.

The waiting lines can be long because it is first-come-first-served. The patient pays for the services received and applies to insurance for reimbursement.

Many Italians take out additional private insurance to improve their situation.

European visitors can present a form E111 to the foreigner's office in exchange for a document covering their temporary stay. Non-EU visitors can obtain private insurance.

Italian physicians are well trained and dedicated. They are well motivated, concerned, and have the respect of the Italian people. Italy has the highest physician-to-population ratio in Europe.

Italy's best private hospitals are as good as any in the world. The state hospitals, with three to six bed wards, could be better.

The Netherlands

Population:	16,491,461
Life expectancy:	78 years
Infant mortality:	5 per 1000 births
Health care:	About 9.8% of GDP

The Netherlands has a very interesting private health care system that covers everyone. Accessibility and quality are under government direction. Residents purchase basic health insurance from one of 93 companies. If a non-resident's stay in the Netherlands is not temporary, purchase of health insurance is required.

Private health suppliers are responsible for the day-to-day management of the health care system. The Dutch government is responsible for the accessibility and quality of the health care.

Features of the Netherlands' health care system include:
- Complete coverage of all residents
- Strong primary care focus: primary caregivers act as gatekeepers; all patients related to one specific practice
- Primary care is offered by trained family medicine specialists
- All other medical specialists work in hospitals (private or salaried)
- Increasing role for nurses.

The basic health care package includes:
- Medical care, including services by general practitioners, medical specialists and obstetricians, and hospitals
- Hospital stay
- Dental care until age 18; those 18 or older are covered only for specialist dental care and false teeth
- Various medical appliances
- Various medicines
- Prenatal care
- Patient transport (ambulance)
- Paramedical care.

The fees for the basic health insurance package, normally about €95 per month, are annually determined by the health insurance companies. The Ministry of Health (Ministerie van Volksgezondheid, Welzijn en Sport) determines the standard premium. The insurance companies determine the additional fee people must pay in the end by charging a certain rate and a no-claim charge. These additional fees form the basis of competition among the various insurance companies. The insurance companies also determine fees and discounts for various good health behaviors. A new law promises to make it easier for consumers to change health insurance companies. Children under 18 are covered for free.

Medical specialists, whether private or contracted, work in hospitals.

Recent changes to the Dutch health care approach have focused on cutting costs. Journalist Gautam Naik reported in the Sept. 7, 2007, *Wall Street Journal*, that, "Since a new system took effect here last year, cost growth is projected to fall in 2007 to about 3% after inflation from 4.5% in 2006. Waiting lists are shrinking, and private health insurers are coming up with innovative ways to care for the sick."

Naik further noted that the new system "hinges on competition among insurers. They are expected to cut premiums, persuade consumers to live healthier lives, and push hospitals to provide better and lower-cost care."

Sweden

Population:	9,000,000
Life expectancy:	80 years
Infant mortality:	3.4 per 1000 births
Health care:	About 8% of GDP

In Sweden's health care system, universal coverage and good health are national goals. The Swedish people believe that health care is a national responsibility and should be paid for from national taxes. The Swedish medical system is excellent, continuously involved in improving access and quality.

The Health and Medical Services Act of 1982 requires that health services of good quality providers shall be offered on equal terms and accessible to all people. The act further stipulates that the services provided shall respect the patient's right to make his own decisions.

The health system is implemented in 21 different areas throughout Sweden by local county councils. These councils operate almost all services and collect taxes to pay for them. Fees are charged to see a primary physician, a specialist or to enter a hospital. Such expenditures are capped, after which all care is free. Private fees amount to about 4 percent of a county council's income.

Sweden is said to have the oldest population in Europe, with 18 percent over age 65. Good health is not uniform throughout Sweden.

Each Swede may choose a family physician and may also choose the hospital, sometimes without a primary referral. Primary care utilizes general practitioners, nurses, and physiotherapists. Primary physicians are about 20 percent of the total doctors. Home care, center care and hospital care are used in Sweden with a growing tendency to favor outpatient care whenever possible. Hospitals specialize. For example, only two well-equipped and staffed hospitals in Sweden perform thoracic and cardiovascular surgery.

Elderly and disabled people can frequently remain in their own homes, because medical services and nursing are available. People in nursing homes and those living in the service apartments have access to nursing services 24 hours a day.

In cases of injury or illness resulting from medical care or treatment, the National Board of Health and Welfare is informed. A patient insurance scheme and county councils make financial awards. In cases of staff fault or negligence, the National Medical Disciplinary Board is informed. The board can decide on disciplinary measures or remove the person from professional duties.It is important to note that staff responsibility and financial compensation are kept separate.

Switzerland

Population:	7,500,000
Life expectancy:	80 years
Infant mortality:	4 per 1000 births
Health care:	About 11% of GDP

The Swiss health care system is the second most expensive system in the world, behind United States.

Swiss health and health care are very good. Public dissatisfaction with the health care system is very low.

Each person must buy compulsory, mandated health insurance from one of the 93 nonprofit insurance entities. Each insurance entity must accept any and all applicants for the compulsory insurance benefit. Health insurance policies may vary, with low deductible or high deductible or HMO structure, with be discounts for good medical behaviors and for non-use. The insurance does not offer full financial coverage.

The price of the insurance is tightly controlled but may vary by canton. Insurance companies are risk-adjusted so that companies with heavier duties receive transfers from those companies with lighter duties.

About one third of the population requires financial support to purchase mandated insurance. This yields about 20 percent of the total premium income. Some cantons pay the financial support to the insurance companies directly and others pay the individual.

This compulsory benefit may be supplemented by additional insurance, usually from a for-profit insurance company that is entitled to refuse coverage for an applicant. Even with the additional coverage, citizens of Switzerland usually have additional expenses that they must pay themselves.

About two thirds of the Swiss consumers' health care costs are for insurance. The other third is for deductibles and non covered items such as drugs and over-the-counter medicines.

The Swiss have a great deal of information about their insurance companies. The insurance companies compete by offering different types of policies with different deductibles. Insurance companies are very interested in and compete for the opinions and wishes of the public because the public is their source of income.

Public information about physicians and hospitals is not readily available. Physicians Associations and the Swiss Insurance Association negotiate fees for services included in the health insurance. Physicians may not supplement their bills but are well compensated.

Study Shows Medicare Improves Health

A team of Harvard Medical School researchers assessed the effect of acquiring Medicare coverage on the health of previously uninsured adults. In the article, "Health of Previously Uninsured Adults After Acquiring Medicare Coverage" (*Journal of the American Medical Association*, Dec. 26, 2007), J. Michael McWilliams, M.D., and his colleagues present the strongest evidence to date that health improves significantly when people gain health insurance. For example, for every 100 uninsured adults with heart disease or diabetes before age 65, the researchers found that with Medicare coverage they had 10 fewer major cardiac complications, such as heart attack or heart failure, than would be expected by age 72.

Authors: J. Michael McWilliams, M.D., Ellen Meara, Ph.D., Alan M. Zaslavsky, Ph.D., John Z. Ayanian, M.D., M.P.P.
Contact: ayanian@hcp.med.harvard.edu
Commonwealth Fund Summary Writer(s): Christopher Gearon and Deborah Lorber.
Source: www.commonwealthfund.org/publications

United States

Population: 200,000,000
Life expectancy: 77 years
Infant mortality: 7 per 1000 births
Health care: About 16% of GDP
Doctor:patient ratio: 2.79 doctors per 1000 inhabitants

We've looked extensively at the United States health care system. Now let's see how it stacks up against health care in other countries. Below is a table that compares our system with others.

Table 1.

Health Care Spending and Physician and Nurse Ratios, 2003

Country	Health Expenditure as Percent of GDP	Total Expenditure on Health Care per Capita (U.S.D PPP)	Public Expenditure on Health Care per Capita (U.S.D PPP)	Private Expenditure on Health Care per Capita (U.S.D PPP)	Physicians per 1000 population	Nurses per 1,000 population
Australia	9.3%	$2,699[a]	$1,821[a]	$878[a]	2.5[a]	10.2
Canada	9.9[c]	3,001[c]	2,098[c]	903[c]	2.1	9.8
Finland	7.4	2,118	1,622	497	2.6	9.3
France	10.1[c]	2,903[c]	2,214[c]	689[c]	3.4	7.3
Germany	11.1	2,996	2,343	653	3.4	9.7
Netherlands	9.8	2,976	1,856	1,119	3.1	12.8[a]
Spain	7.7	1,835	1,306	529	3.2	7.5
Sweden	9.4	2,703	2,304	399	3.3	10.2a
United Kingdom	7.7[a]	2,231[a]	1,860[a]	371[a]	2.2	9.1
United States	15.0	5,635	2,503	3,131	2.3[a]	7.9[a]

Source: Organization for Economic Cooperation and Development (OECD) Health Statistics, 2005.
Note: PPP = Purchasing power parity (an estimate of the exchange rate required to equalize the purchasing power of different currencies, given the prices of goods and services in the countries concerned.)
   a = 2002 data
   c = Estimate

EXHIBIT 1

List of 32 Indicators in State Scorecard on Health System Performance

Access	Year	All States Median	Range of State Performance (Bottom–Top)	Top State
1. Adults under age 65 insured	2004–2005	81.5	69.6 – 89.0	MN
2. Children insured	2004–2005	91.1	79.8 – 94.9	VT
3. Adults visited a doctor in past two years	2000	83.4	73.9 – 91.5	DC
4. Adults without a time when they needed to see a doctor but could not because of cost	2004	87.2	80.1 – 96.6	HI
Quality				
5. Adults age 50 and older received recommended screening and preventive care	2004	39.7	32.6 – 50.1	MN
6. Adult diabetics received recommended preventive care	2004	42.4	28.7 – 65.4	HI
7. Children ages 19–35 months received all recommended doses of five key vaccines	2005	81.6	66.7 – 93.5	MA
8. Children with both medical and dental preventive care visits	2003	59.2	45.7 – 74.9	MA
9. Children with emotional, behavioral, or developmental problems received mental health care	2003	61.9	43.4 – 77.2	WY
10. Hospitalized patients received recommended care for acute myocardial infarction, congestive heart failure, and pneumonia	2004	83.4	79.0 – 88.4	RI
11. Surgical patients received appropriate timing of antibiotics to prevent infections	2004–2005	69.5	50.0 – 90.0	CT
12. Adults with a usual source of care	2004	81.1	66.3 – 89.4	DE
13. Children with a medical home	2003	47.6	33.8 – 61.0	NH
14. Heart failure patients given written instructions at discharge	2004–2005	49	14 – 67	NJ
15. Medicare patients whose health care provider always listens, explains, shows respect, and spends enough time with them	2003	68.7	63.1 – 74.9	VT
16. Medicare patients giving a best rating for health care received	2003	70.2	61.2 – 74.4	MT
17. High-risk nursing home residents with pressure sores	2004	13.2	19.3 – 7.6	ND
18. Nursing home residents who were physically restrained	2004	6.2	15.9 – 1.9	NE
Potentially Avoidable Use of Hospitals & Costs of Care				
19. Hospital admissions for pediatric asthma per 100,000 children	2002	176.7	314.2 – 54.9	VT
20. Asthmatics with an emergency room or urgent care visit	2001–2004	15.5	29.4 – 9.1	IA
21. Medicare hospital admissions for ambulatory care sensitive conditions per 100,000 beneficiaries	2003	7,278	11,537 – 4,069	HI
22. Medicare 30-day hospital readmission rates	2003	17.6	23.8 – 13.2	ID
23. Long-stay nursing home residents with a hospital admission	2000	16.1	24.9 – 8.3	UT
24. Nursing home residents with a hospital readmission within three months	2000	11.7	17.5 – 6.7	OR
25. Home health patients with a hospital admission	2004	26.9	46.4 – 18.3	UT
26. Total single premium per enrolled employee at private-sector establishments that offer health insurance	2004	$3,706	$4,379 – 3,084	UT
27. Total Medicare (Parts A & B) reimbursements per enrollee	2003	$6,070	$8,076 – 4,530	HI
Healthy Lives				
28. Mortality amenable to health care, deaths per 100,000 population	2002	96.9	160.0 – 70.2	MN
29. Infant mortality, deaths per 1,000 live births	2002	7.1	11.0 – 4.3	ME
30. Breast cancer deaths per 100,000 female population	2002	25.3	34.1 – 16.2	HI
31. Colorectal cancer deaths per 100,000 population	2002	20.0	24.6 – 15.3	UT
32. Adults under age 65 limited in any activities because of physical, mental, or emotional problems	2004	15.3	22.8 – 10.8	DC

Note: All values are expressed as percentages unless labeled otherwise. See Appendices B1 and B2 for data source and definition of each indicator.

SOURCE: Commonwealth Fund State Scorecard on Health System Performance, 2007

CHAPTER 12

States' Efforts to Provide Health Care

Several states are exploring or experimenting with the possibility of statewide, state run basic health care for all residents of those states. These experiments in statewide health care signify and demonstrate urgency. A successful experiment might be extrapolated into a national plan.

State of Maine

Maine has suffered a bigger hit from the increasing health care/gross state product ratio than the national average.

In addition to insurance premiums, deductibles and co-pays, Mainers pay for health care through taxes. Even if they don't have health insurance themselves, they're still paying for someone else's health care through taxes.

Here are some of the many ways Maine pays for health care:
- Insurance premiums
- Out-of-pocket expenses
- Federal income taxes
- Federal income taxes, which pay for Medicare and Medicaid
- Federal taxes also pay for the insurance premiums of federal employees, elected officials, and military employees, as well as the health care benefits of veterans
- State income taxes; a portion of which also pays for Medicare and Medicaid

- Sales taxes, which amounted to 48% of tax revenue in 2004
- Death, gift, documentary and stock transfer taxes, all of which help fund the same health care costs as sales tax dollars
- Cigarette taxes, which primarily fund health care, according to the Government Performance Project which ranks states on their fiscal viability
- Excise taxes which fund town and county employees' health insurance premiums
- Property taxes which fund insurance plans of everyone from sheriff's deputies to town clerks.
- Hunting and fishing licenses which go into the general fund, that in part, finances health care, paying health insurance premiums for state employees and funding the state's portion of Medicare and Medicaid.
- Workers compensation premiums which compensate workplace-related illnesses and injuries. According to Insurance Strategies, Ltd., an insurance industry think tank, approximately 50 percent of workers compensation pays for health care.

The Solution

Maine's solution to health care for all its citizens is the LaMarche Health Care Plan. The Plan also calls for the creation of graduate schools to train medical doctors, dentists and pharmacists to alleviate the shortage of providers in Maine.

The LaMarche Health Care Plan springs out of recommendations for health care coverage in the 171-page Maine Health Care Reform Commission (MHCRC) report released in November 1995.

The MHCRC recommendations spell out the basic health coverage parameters for all Mainers.

- One pair of eyeglasses every three years, or as the patient's prescription changes
- Any doctor or provider of the patient's choice for preventative care, wellness and illness
- Independent health care providers who will bill the Maine Health Care Authority for services rendered
- Prescriptions--non-elective prescription medications
- Braces and pediatric dentistry
- Portable across state lines. If the best care available for a patient

is out-of-state, the plan will pay for that care

- Every Maine resident is included; there's a one-year waiting period for those who move to Maine without immediate employment or educational enrollment
- Maine veterans can go to any Maine hospital
- Home health care services included when prescribed by a provider
- Elective procedures are not included or covered

The LaMarche Health Care Plan calls for the creation of the Maine Health Care Authority to administer the program. The MHCRC recommended that the Authority board be comprised of health care providers, government appointees, hospital administrators and representatives of the business community as well as private citizens. The Authority will oversee the health care program as well as steward the education of health care professionals.

The LaMarche Administration will petition the release of federal Medicaid funds, in accordance with Section 1915(b) of the U.S. Social Security Act, to the State of Maine for incorporation into the plan.

The federal government will be billed directly for expenses related to federal employees living in Maine. The state already employs this direct billing for unemployment compensation paid to federal employees. "Federal employees" are defined as current and retired federal employees, as well as military personnel.

Private employers will be taxed based on two factors: wages paid to employees and the size of their business. Based on the number of their employees, employers will pay between 5 percent and 12 percent of the employees' wages to the state of Maine. These funds will then go directly to the Maine Health Care Authority.

Health Care Cost Basis for Private Employers

Number of employees	Percent of payroll
5 or fewer	5%
6-24	6%
25-50	7%
51-100	8%
101-250	8%
251-500	10%
501-1000	11%
over 1000	12%

For most businesses, the amount paid under the LaMarche Health Care Plan will represent an enormous savings compared to their current expenditures for employee health care premiums. Employers who do not currently provide health insurance for their employees might spend more. However, those employers may, if they choose, adjust wages to compensate for this increased benefit and thereby incur no new net expenses under LaMarche. The only businesses potentially impacted under this plan are those that pay minimum wage and do not provide health insurance, since they cannot adjust wages downward.

Surplus funds generated by the universal system will accumulate in a rainy day fund administered by the Maine Health Care Authority. When this fund reaches $500 million, the percentages assessed for payroll contribution will be adjusted downward.

Anticipated Benefits of LaMarche Health Care Plan

The LaMarche Health Care Plan will eliminate workers compensation medical costs incurred by businesses.

The LaMarche Health Care Plan will eliminate the cost shifting caused by the uninsured.

Health care-related bankruptcies will be substantially reduced. These bankruptcies happen to people who have health insurance, then become seriously ill and lose the jobs that provided the insurance. They lose their jobs, their benefits and their ability to pay.

The LaMarche Health Care Plan will eliminate lost revenue and expenses due to unpaid bills and collection expenses. Funds will be used to pay for health care and not for insurance.

Administrative costs will be drastically reduced; billing will be simplified providers will submit bills to a single payer, the Maine Health Care Authority.

The LaMarche Health Care Plan calls for the establishment of graduate schools in Maine to train medical doctors, dentists and pharmacists.

New health care schools will create jobs and brings millions of dollars to Maine. These schools employ workers at all levels, from professors and administrators to cafeteria workers.

State of Vermont

The Health Care Affordability Act, H.861 Committee of Conference—May 5, 2006.

The first change in Vermont's health care system is shifting the focus, from treating acute illness to focusing on managing chronic diseases. This will slow the rate of growth in health care costs.

A new program will provide affordable, comprehensive coverage for uninsured Vermonters.

Vermont's health care system focuses on the treatment of acute illnesses (conditions that last a short time and can usually be cured). Over the past 50 years, chronic conditions have represented an increasing share of health care costs. Today, 75 percent of all health care spending is on care for people with chronic conditions, yet only 55 percent of this is the right care at the right time. If chronic conditions are well managed, the need for expensive care such as hospitalization is reduced

There will be a new insurance market for people without insurance, or without adequate insurance from an employer. Insurers will be invited to offer in this market. There will be one standard plan. It will look a lot like the typical insurance plan that is offered in Vermont today, with one major difference: there will be no cost to the patient for preventive care such as mammograms or for recommended services for chronic illness, such as eye exams for people with diabetes.

People who are uninsured and not eligible for adequate coverage through employment will be eligible to purchase Catamount Health. Anyone under 300 percent of poverty (about $60,000 for a family of four) will receive assistance with their premiums from the state. Uninsured means that the individual either has been without coverage for 12 months, or has lost coverage for reasons such as job loss, divorce, dissolution of a civil union, or death of the primary policy holder.

The second major part of Catamount Health is a mechanism to provide coverage for people who are uninsured, but eligible for insurance through their employers, if the insurance meets coverage standards. In this case, the state will help with the employee share of the premiums and with cost sharing (deductibles, coinsurance) for care related to chronic conditions.

The state will provide assistance with premiums, based on family income. Second, payments to providers under Catamount Health will be less than private insurers pay, but more than Medicaid and Medicare, so Catamount Health will not contribute to the cost shift. In fact, because people who previously had no insurance will now be covered, the cost shift will be reduced (less bad debt and free care).

The state will apply to the federal government for permission to include Catamount Health in Vermont's Medicaid waiver. If permission is granted, that will mean that the federal government will pay about 60 percent of the cost of Catamount Health. The state's share will come from two increases in the cigarette tax, from $1.19 to $1.79 next year, and to $1.99 in 2009 and from an assessment on employers for employees who either are not offered insurance or who are offered insurance, chose not to enroll, and are uninsured.

State of Massachusetts

In April 2006 the Massachusetts Legislature created a statewide mandated, private, universal health care plan. Since the implementation of Massachusetts' landmark Healthcare Reform began in June 2006, an estimated 290,000 residents of the Commonwealth have become newly insured.

The Massachusetts Health Connector is the central purchasing pool through which insurance is offered at lower than average rates. The Commonwealth Care Health Insurance Program (Commonwealth Care) is run by the Health Connector. It connects eligible Massachusetts residents with approved health plans and helps them pay for them. All residents are required to get health insurance or face a fine or tax penalty.

Businesses establish Section 125 accounts. Employees pay for insurance on a pretax basis.

Subsidies for families below the 300 percent of the poverty line. Earnings between $30,000 and $50,000 are not be eligible for state subsidies. But they aren't be penalized if they cannot find health insurance and cost between $150 and $300 a month. Waivers may be available for people who genuinely cannot afford coverage.

Limited insurance for individuals above 300 percent of the poverty line.

Financing is in part through federal funds and state funds from the uncompensated care pool. Massachusetts has a state uncompensated care pool in excess of $500 million. This came from a law allowing hospitals to bill the state for treating low income patients.

Massachusetts was unique among states in three ways:

1 Fewer uninsured: Massachusetts = 9 percent vs. 18 percent nationally
2 Mandated universal insurance purchase. Everyone is covered
3 Medicaid section 1115 waiver was required by CMS to increase insurance coverage or be lost.

By January 1, 2009, residents are required to have a plan that offers some key benefits. These benefits are what the Health Care Reform Law calls "Minimum Creditable Coverage." These benefits are set by the Health Connector. They are the minimum benefits expected for Massachusetts adults. They include:

1 Prescription drug coverage
2 Visits to the doctor for preventative care, before a deductible.
3 Deductibles that are capped at $2,000 for an individual or $4,000 for a family each year
4 An annual cap on out-of-pocket spending at $5,000 for an individual or $10,000 for a family (for plans with up-front-deductibles or co-insurance)
5 No cap on total benefits for a particular sickness or for a single year
6 No cap on payment toward a day in the hospital.

Critics claim that it will be expensive for family coverage. There is a sliding scale of affordability standards. The highest premium is $8,640 per year for families with children earning $90,000 to $110,000 per year. The cheapest rate is $420 per year for singles earning $15,000 to $20,000 per year.

Employers' relief from offering health care is a savings for employers that should be shared in order to diminish costs. It will require a great deal of compromise and sacrifice to make this work but Massachusetts deserves a lot of credit.

State of Illinois

Governor Rod R. Blagojevich is pushing a new health care initiative that would offer insurance and wellness training to everyone in Illinois in a plan called Illinois Covered. Its proposed financing is via a tax on gross business receipts. The general assembly is opposed to the new tax, as are other interests in the state. It is too early to know the outcome.

State of California

California has been diligent in attempting universal health care.

In 2005-6 California passed Senate Bill 840 providing for universal, single-payer health care coverage. A well thought out project, it was vetoed by Governor Arnold Schwarzenegger.

In January 2007, Gov. Schwarzenegger proposed a universal, mandated, individually owned health care policy for everyone in the state of California. It would cover the 6.7 million uninsured Californians.

The Los Angeles Times reported on February 6 that "some of the biggest players in [California]'s health care industry have agreed to commit millions of dollars to a campaign for universal health care access." The New America Foundation has worked extensively with Gov. Schwarzenegger and other groups in California to craft this health reform plan, which aims to provide coverage to all Californians and create a system with truly shared responsibility.

The *Times* reported that "the yet-unnamed alliance... includes a labor giant, the Service Employees International Union; the state's largest doctors' lobby, the California Medical Association; the state's biggest nonprofit hospital chain, Catholic Health Care West; and three major insurers: Kaiser Permanente, Blue Shield of California and Health Net."

A coalition of 36 companies is planning a campaign to expand medical insurance nationwide in the manner similar to the California plan. The backers included PepsiCo, General Mills, Aetna, Blue Shield of California, PacifiCare, Eli Lilly and CIGNA Health care.

While the Democrat-controlled California Assembly passed the plan in December 2007, state senators and Governor Schwarzenegger have yet to reach agreement. There is concern about the plans' $14.4 billion price tag when the state faces a budget deficit of about the same size, according to the December 25, 20007, *New York Times*.

As in Massachusetts, the California plan would mandate coverage for most individuals. It would raise money to subsidize policies for low-income residents through what Mr. Schwarzenegger calls shared responsibility—a tax on hospital revenues, a hefty increase in tobacco taxes and assessments on employers who do not contribute to their workers' health care, according to the *New York Times*.

State of Colorado

The Colorado Health Foundation is initiating a project to make Colorado the healthiest state in America. The foundation's strategy is "to influence health policy by developing informed, engaged and passionate leaders, by bringing together people in organizations with a common agenda, and by strengthening advocacy groups in helping the public better understand health policy issues." To that end, the Colorado Health Foundation issues an annual Colorado health report card measuring the health of children and adults in Colorado, and tracking health care progress.

"Crank? a man with a new idea until it succeeds."

Mark Twain

CHAPTER 13

Other Proposals

Acommon dilemma is faced by all office holders and political candidates who present health care plans. The dilemma is this. At present, employers provide about 60 percent of workers' health care. If a reform plan relieves the employer of any and all health care obligation, what happens to the money that paid for that care?

A reform that co-opts all the money is of no benefit to the employer, who truly needs a benefit.

A reform that takes none of the employer money will be very expensive.

Another possible reform approach would be to release the employers who offer health care and tax all employers a percentage amount that is less than those offering health care are paying now. That would benefit employers and still raise some of the money to pay for health care.

It's interesting to see how each candidate and political office holder handles this issue. The proposed reforms attempt to take a middle-of-the-road position.

With regard to the 2008 presidential candidates, a website to check is: http://www.health08.org. This website, sponsored by the Henry Kaiser Family Foundation, lets you compare the various health care plans side by side.

Senator Hillary Clinton
(www.hillaryclinton.com)

Sen. Clinton's initial plan would:
- Cut health care costs (i.e., better management of chronic care)
- Improve quality (i.e., electronic medical records and information

technology)
- Cover everybody (requires canceling some tax cuts)
- Achieve political will by creating a broad coalition of business, labor, physicians, nurses and hospitals to withstand resistance to reform from insurance and pharmaceutical industries.
- Caps the percentage of annual income (as yet unspecified) that can be mandated for purchase of health care insurance.

On September 17, 2007, Senator Clinton proposed a national health insurance program, requiring everyone to purchase health insurance. Federal subsidies would make the insurance cheaper, and tax credits would help the people at the bottom of the income scale. The American Health Choices Plan would allow consumers to keep coverage that they already have, if they so desire. There would be a choice of private plans to purchase, similar to the choices that members of Congress receive, and a public plan option similar to Medicare.

Senator Clinton emphasizes the moral imperative to ensure that "every single American has quality affordable health coverage."

The plan's advantages are:
- It would cover all of the 47 million uninsured people.
- Its estimated cost: $110 billion.
- Working families receive a refundable tax credit. This might serve as an incentive for the government to help lower income families to increase their income so that less government financial help will be necessary.
- Choice of insurance company and plan is allowed.
- Private insurance companies have a large role.
- Various players are expected to cooperate or partner to improve the situation.
- No pre-existing conditions are allowed.
- The use of health information technology is promoted.
- Premium payments are limited to a percentage of income
- Generic drugs are mandated, at low cost
- A public-private institute will evaluate new drugs, devices and treatments
- All who benefit from the system play a part in its success
- Large employers must provide health insurance or contribute to the cost of coverage
- Quality and modernization produce savings

- The gold plated plans are trimmed, but employers retain their tax exclusion for health care
- Discontinues portions of the Bush tax cuts for those making over $250,000 a year
- It's a relatively simple plan.

Senator Clinton's plan's disadvantages are:
- Perception of big, expensive government.
- Choice of plans: If the consumer can initially pick an inexpensive plan with restricted benefits, will he/she later be able to buy a plan with adequate benefits? If so, this could bankrupt the insurance companies. If not, how will the care that later becomes expensive be delivered? A choice of plans increases management overhead for insurance companies.
- Changing to another company: Will the consumer be able to switch to another company? This should allowed if it's perceived that another insurer offers better service and better value. A company that does an excellent job could end up with all the expensive cases. Such a good performing company needs protection such as with reinsurance.
- A public plan option like Medicare would be very difficult to compete against. Its overhead would be lower than the private sector with its multiple plans to be managed.
- Possible increased costs for everyone.
- Mandates are unacceptable to some people as interfering with individual freedom.
- It does not replace Medicaid or SCHIP.
- Large employers will be expected to provide health insurance or contribute to the cost of coverage.

Senator Barack Obama (www.barackobama.com)

Sen. Obama's health care plan would:
1 Levy a tax on employers who do not offer employer-sponsored health insurance, hoping to induce employers to offer it. If not, the tax will be used for expanded public health programs intended to subsidize affordable health insurance for the uninsured working poor.

2   It provides the Federal Employees Health Benefit Program to everybody, with subsidies for the working poor.

3   Insurance companies may not deny coverage or charge higher premiums based on medical history.

4   It organizes health exchanges to serve small companies and individuals by collecting bids and assuring quality from insurance companies to offer family choices.

5   Purchase of health care is not mandated. Sen. Obama believes that most people want health care but cannot afford it. He believes they will step up and purchase health care if it is available at a reasonable price. He believes that his plan will save about $1000 a year for the average family.

6   Children's coverage is mandated.

7   If these steps fail and/or if people refuse in large numbers mandated coverage could follow.

This plan will appeal very much to people who know their care will be expensive and they'll definitely participate.

However, denying insurance companies coverage any options or higher premiums based on medical history can be fair and workable only if everyone participates, including all the healthy people. Mandated coverage is the flip side of guaranteed acceptance.

Relying on voluntary participation to achieve universality is challenged by other similar experiences. For example, participation in voting, which is a free civic duty, is considerably less than universal. Voluntary participation in intelligent personal health practices is less than universal.

Fortunately or unfortunately, people who do not participate would have to receive care anyway in emergencies, thus driving up the costs for those who do participate.

Senator John Edwards
(www.johnedwards.com)

The Edwards plan achieves universal coverage by:

- Requiring businesses and other employers to either cover their employees or help finance their health insurance. Total costs estimated at $90 to $120 billion would come from repeal of tax cuts on incomes over $200,000 a year.
- Making insurance affordable by creating new tax credits, expanding

Medicaid and SCHIP, reforming insurance laws, and taking innovative steps to contain health care costs (information technology, increase efficiency, no pre-existing conditions and covering mental health)

- Creating regional Health Care Markets to let every American share the bargaining power to purchase an affordable, high-quality health plan, increase choices among insurance plans, and cut costs for businesses offering insurance.

Once these steps have been taken, requiring all American residents to get insurance. Children's coverage is mandated.

It seems there is nothing impossible about Sen. Edwards' plan but it will require time and information technology.

Comparing three candidates' positions on health care:

Estimated annual costs (all call for rolling back President Bush's tax cuts):

Clinton	$110 billion
Obama	$50 billion to $65 billion
Edwards	$90 billion $120 billion

Universal coverage:

Clinton	Individual mandate
Obama	*Make It Affordable*, mandates for children
Edwards	Individual and small business mandate

Choice of Plans:

Clinton	Keep your own, private plans, Medicare-like
Obama	Plans similar to federal employees
Edwards	Choice of plans with tax credits

Employers:

Clinton	Large employers play or pay
Obama	Large employers play or pay
Edwards	Employers play or pay partnership

Representative Dennis Kucinich (www.dennis4president.com)

Rep. Kucinich's health care plan is the single-payer system. He has, together with Representative John Conyers, introduced a bill in

Congress, HR 676, to achieve universal coverage with a single-payer system. It is supported by 19,000 physicians.
- All Americans citizens receive benefits
- Primary and preventive care
- Inpatient and outpatient services
- ER services
- Prescription drugs
- Long-term care
- Mental health services
- Dental coverage
- Hearing and vision coverage
- Substance abuse treatments
- Chiropractic services
- No deductibles, no co-pays, no out-of-pocket expenses
- Portable, see any doctor in United States.

Senator John McCain
(www.johnmccain.com)

On October 10, 2007, Sen. McCain unveiled a plan to contain spending by emphasizing better treatment of chronic diseases such as diabetes and heart disease, which account for 75 percent of health care costs. He also wants accountability from drug companies, insurance companies, doctors, hospitals, the government and patients. He would achieve this by linking compensation to performance. He proposes a single charge for high-quality care. He would give a tax credit of $2500-$5000 to low income families who obtain their own insurance.

Employers would no longer be allowed to deduct health care costs from their taxes. He would not require insurance purchase. He would work to lower drug costs by introducing generic drugs and enabling re-importation.

Rudolph W. Giuliani
(www.joinrudy2008.com)

Mr. Giuliani proposes an income tax credit of up to $7,500 for individuals and $15,000 for families to help pay for health insurance of their choice. He would improve health savings accounts. He favors tax cuts, increasing competition and empowering patients and doctors.

Mitt Romney
(www.mittromney.com)

Mr. Romney proposes that the 50 states devise their own health care plans. He proposes deregulation of health insurance markets, to encourage cheaper policies.

Mitt Romney would allow tax deductions for premiums, deductibles and co-payments. He is opposed to mandates and hopes for coverage by driving down costs with market reforms. He would assist low income Americans in buying private health insurance. He has confidence in the free market approach. He wants less government, particularly in health care insurance with easing of regulations.

And Other Health Care Plans and Proposals...

The urgent need for change in U.S. health care insurance is best observed by noticing how many different organizations and governmental entities are making proposals. We've already seen that political candidates and officials realize health care is something that they must address. The states are attempting to enact legislation of various types to the issue. Here is a potpourri of more proposals to fix our clearly inadequate if not broken health care insurance system.

Federation of American Hospitals
(www.fah.org)

In February 2007, the Federation of American Hospitals proposed a health care plan which would require individuals to take coverage through employers when health benefits are offered or purchase it on their own or to receive it through existing government programs. It would be a means tested program. This would cost $115 billion to $900 billion that federal and state agencies currently expend on health care. The hospitals of America would like to be relieved of a $40 billion a year subsidy to people without insurance.

Mayo Clinic Plan
(www.mayoclinic.org/healthpolicycenter/recommendations.html)

On September 15, 2007, the Mayo Clinic proposed mandated private health care plans. Insurance companies would offer standard plans and applicants could not be turned down. The Mayo Clinic recognized the diminishing and eventual ending of employer health plans. But under the Mayo plan, employers still have a health care role and contribute to financial help . Employees would be able to keep their individual policies when they change jobs. Under the Mayo Plan, the federal government would subsidize lower income people, because insurance now costs $11,000 or $12,000 a year.

City of San Francisco (www.healthysanfrancisco.org)

Healthy San Francisco is the first city health care program to guarantee care to all of its uninsured. Financed by the city, San Francisco legislators believe universal care to the uninsured can be achieved using the same money now being spent on emergency room treatments. In September 2007, the program began covering the entire city with services at 22 city-based clinics. San Francisco's population of 750,000 includes a relatively small number of uninsured adults who now are welcome to the clinic services. Future plans include bringing private medical networks into the program to offer wider choices of doctors. After the first three months of limited enrollment, coverage expanded to any San Francisco resident who has been uninsured for 90 days. The plan is not an insurance and is not portable, the Healthy San Francisco website advises, so if one has health insurance, do not drop it; it is the better choice.

Michael Levitt, U.S. Secretary of Health and Human Services

Secretary Michael Levitt proposes a value driven health system characterized by better care at lower-cost. "As value in health care becomes transparent, everything improves: costs stabilize; more people are insured; more people get better health care; and economic competitiveness is preserved," he says. "This is a prescription for a value driven system— a prescription of good medicine networks for everyone."

Wal-Mart Plan

As the nation's largest retail employer, Wal-Mart has come under fire about health insurance. Websites have sprung up taking the retail giant to task for many issues, including how it goes about insuring its workers' health. So for 2008, Wal-Mart has unveiled new improved health insurance options for its full and part time workers. The menu of insurance includes choices such as $100 to $500 in pre-deductible health care credits, various insurance plan options including one with a $5 per month premium without hospital deductibles. Other deductibles could reach $2000 per year. There would be a waiting period of one year. Generic prescriptions would cost $4.

Critics of Wal-Mart say that, for the vast majority of Wal-Mart workers, who make an average annual salary of $18,800, these plans are still unaffordable due to low wages or inaccessible due to waiting periods, said to be twice the national retail average. With Wal-Mart's continuing trend toward more part-time employees, who will not be eligible for health care coverage until they have been with Wal-Mart for one year, these changes may do little to provide coverage for Wal-Mart's average employees.

As of January 2007, 9.6 percent of Wal-Mart's 1.3 million workers had no health insurance of any kind. It is unclear how many can and will sign up under the new plan.

http://www.usatoday.com/money/industries/retail/2007-01-11-%20walmart%20insurance_x.htm

"You can always spot a well-informed man—

his views are the same as yours."

                     Ilka Chase

CHAPTER 14

How to Keep Up

It's an ever-changing scene, with new plans being proposed, and discussion raging fiercely. How can you stay abreast of it all? The best resources are available on the internet, where updates are frequent. Not all websites are equal, however. Here are some online resources that offer reliable information.

Outstanding health care blog sites:

- Health Care Policy and Marketplace Review: latest developments
  www.Healthpolicyandmarket.blogspot.com/search/
- Joe Paduda's Managed Care Matters
  www.joepaduda.com
- Health Affairs' Blog The flagship of all health care blogs
  http://Healthaffairs.org/blog
- The Best WC Site "Workers Comp Insider"
  www.workerscompinsider.com
- Richard Eskow's "The Sentinel Effect"
  http://sentineleffect.wordpress.com
- Matt Holt's "The Health Care Blog"
  www.thehealth careblog.com/
- Lindsay Resnick's "Resnick Unplugged"
  www.lindsayresnick.com
- David Harlow's HealthBlawg
  http://healthblawg.typepad.com/
- Insurance Topics at Insureblog
  http://insureblog.blogspot.com/

- David Williams' Health Business Blog
  www.healthbusinessblog.com/
- Jason Shafrin's Health Care Economist
  http://health care-economist.com
- Voice for the Uninsured
  http://www.ama-assn.org/ama/pub/category/17712.html
- Families USA : The voice for health care consumers
  http://www.familiesusa.org
- Brian Klepper's page of resources
  www.brianklepper.net
- The Henry Kaiser Family Foundation website tracking the various 2008 presidential candidates' health plans:
  www.health08.org

State websites that report quality data:

- California: www.calhospitalcompare.org

- Colorado: www.hospitalquality.org

- Florida: www.floridacomparecare.gov;
  www.floridainformedpatient.com

- Georgia: http://gahospitalqualitycheck.org

- Iowa: www.ihconline.org/iowareport/iowareport.cfm

- Kentucky: www.kyha.com/QualityData

- Massachusetts: www.patientsfirstma.org

- Michigan: www.mha.org/mha_app/keystone/index.jsp

- Minnesota: www.mnhospitalquality.org

- Missouri: www.focUS.onhospitals.com

- New Hampshire: www.nhqualitycare.org

- New Mexico: www.nmhhsa.org

- North Carolina: www.NCHospitalQuality.org

- New York: http://www.health.state.ny.US.

- Ohio: www.ohanet.org/portal/default.htm

- Oregon: www.orhospitalquality.org

- Pennsylvania: www.phc4.org

- Rhode Island: http://hari.org/quality.shtml

- Utah: www.utcheckpoint.org

- Vermont: www.vahhs.org/Act53/Act53Main.htm

- Washington: www.wahospitalquality.org

- Wisconsin: www.wicheckpoint.org

- Top 100 Hospitals: This site compares hospitals on 19 quality measurements, not outcomes. This is a very important site which allows meaningful comparisons amongst hospitals. www.hospitalcompare.hhs.gov

References

- California Healthcare Foundation.
- "Lifestyle Is the Primary Determinant in the Top Six Leading Causes of Death in the U.S." National Center for Health Statistics. 2002.
- Budnitz, Daniel, M.D., *National Surveillance Project on Outpatient Drug Safety.* The Centers for Disease Control and Prevention, Food and Drug Administration and U.S. Consumer Product Safety Commission.
- "Levels of Dissatisfaction with the U.S. Health System Vary Sharply between the Insured, and Those Who Have No Health Insurance." *USA Today*/ABC News/Kaiser Family Foundation poll. September 7-12, 2006.
- N.C.H.C.
- "More Americans Head Overseas for Health Care." Associated Press. Nov. 02, 2006.
- Gibbons, Raymond, M.D., President, American Heart Association. "Fundamental Change in Health Care Delivery." November 2006.
- Gale, Arthur H., M. D., "AMA Restores Hippocratic Oath to Medical Ethics," *Missouri Medicine.* Vol. 98: No. 20, pages 532, 533, 534. December 2001
- Harris, Michelle, "In-Store Medical Clinics Becoming a Growing Trend."
- "The cost shift payment hydraulic." *Health Affairs.* January 2006
- National Federation of Independent Business. Gallup Organization.
- Appleby, Julie, "Wal-Mart Memo Sparks Criticism." *USA Today*, October 27, 2005.
- Kaiser Family Foundation.
- Klepper, Brian, and Salbur, Patricia, "The Business Case for Reform." *Modern Healthcare.* October 10, 2005.
- Shea. (Power Point)
- "Part D. Enrolling Well, Costing More." America's Pharma and Healthcare Insight. September 2006.
- *Consensus in America: Affordable Health Care for All.* Citizens Health Care Working Group. September 25, 2006.

- The American Journal of Health Promotion.
- Freudenheim, Milt, "Hospital Group Offers Plan on Health Coverage for All." *New York Times.* February 22, 2007.
- "Money and Markets." Weiss Research, Inc., October 9, 2006.
- Leonhardt, David, "A Lesson from Europe on Getting Better Health Care for Less." *New York Times.*
- Appleby, Julie. "The U.S. Health Care System," *USA Today.* Oct. 16, 2006.
- Press release, Agency for Healthcare Research and Quality. August 11, 2005.
- Klepper, Brian. Personal communication. November 10, 2005.
- "What's behind the rise in health care costs?" News Max.com/e-mail March 6, 2007.
- "Health Insurance Cost," National Coalition on Health Care. March 6, 2007.
- Mayo Clinic Facts. 2004.
- Leonhardt, David. "The Data Tell a Different Story on Heart Patients." *New York Times,* March 7, 2007.
- Toner, Robin, and Elder, Janet, "Most Support U.S. Guarantee of Health Care." *New York Times,* March 2, 2007.
- Haige, Scott, M. D. "Doctors Without Dollars." *Time,* February 12, 2007.
- Executive Summary, National Symposium on Health Care Reform.
- The Massachusetts Health Care Reform. *The Hastings Center Report.* Vol. 36, No. 5, September-October 2006, pages 14–29.
- Owcharenko, Nina, and Moffit, Robert, "Backgrounder." The Heritage Foundation. No. 1953. July 18, 2006.
- California Senate Bill No. 840, amended August 24, 2006: Single-Payer Health Care Coverage.
- "Together we can make Colorado the healthiest state in the nation." The Colorado Health Foundation Report. 2006.
- "The Colorado Health Report Card." The Colorado Health Foundation Report. 2006.
- Ostergren, Allan, "A Universal Tax Credit" *New York Times Letters to the Editor,* March 8, 2007.
- "Using Blunt Force on Missouri's Most Vulnerable Population," Publication No. 07-103. Families USA, Washington, DC. 2007

- Gale, Arthur H., M.D., "AMA Restores Hippocratic Oath to Medical Ethics," *Missouri Medicine*, Vol. 98: No. 20, pages 532, 533,534.
- Saulny, Susan, "Tax to Pay for Health Plan and Illinois Faces Resistance." *New York Times*, May 5, 2007.
- Stamm, Rebecca, "Death of the Health Insurance Industry."
- Arthur H. Gale M. D., "AMA Restores Hippocratic Oath to Medical Ethics," Missouri Medicine, Vol. 98: No. 20, pages 532, 533, 534. December 2001.
- Freudenheim, Milt, "Hospital Group Offers Plan on Health Coverage for All." New York Times, February 22, 2007.
- Belluck, Pam, "Massachusetts Agency Proposes Insurance Costs That Most Can Afford." New York Times, April 12, 2007.
- Miranda, Hitti, "25 Steps to Better Health." CBS News. 2006.
- Davis, David A., M.D., et al. "Accuracy of Physician Self-Assessment Compared with the Observed Measures of Competence." Journal of the American Medical Association. September 6, 2006, p. 1094.
- Heneghan, Kathleen, R.N.; Sachdeva, Ajit K., M.D.; Aninch, Jack W., M.D., "Transformation to a System That Supports Full Patient Participation." *Bulletin of the American College of Surgeons*. Vol. 91: No. 6, page 12.
- Rogers, Arvey I., "The Cornerstone of Medicine: The Physician-Patient Relationship." *Miami Medicine*, Dade County Medical Association. November 2006.
- Kalb, Claudia, and Murray, Christopher, "Eat Your Veggies." *Newsweek*, page 8, September 25, 2006.
- Rudd, Rima, M.D., "Most Adults Have an Intermediate Health Literacy." The National Center for Adult Learning and Literacy.
- Gingrich, Newt, "Saving Lives and Saving Money: Transforming Health Care in the 21st Century." The Center for Health Transformation. 2005.
- The American College of Medical Quality.
- President's Commission for the Study of Ethical Problems in Medicine and Biomedical and Behavioral Research. 1983.
- "Protecting Research Subjects with Diminished Capacity." The Science and Ethics Literacy Project. Issue 2,. Summer 2006.
- Bacon, John, "Big Business Warms to Healthcare Changes." USA Today, April 24, 2007.

- Chantler, Cyril, M.D., et al., "Information Technology in the English National Health Service." *Journal of the American Medical Association.* Vol. 296: No. 18, p. 2255. November 8, 2006.
- "The Next Generation of Healthcare." Alegent Health.
- Schoen, Cathy, et al., "U.S. Health System Performance: A National Scorecard." *Health Affairs*, September 20, 2006, p. 457.
- Willshire, Ronda, "Jumpstarting Health Care Reform." *Mayo Clinic Magazine.* Summer 2006.
- "Specialty Hospitals" report, U.S. Government Accounting Office. October 2003.

About the Author

Edwin Tutt Long, M.D.

A board-certified thoracic surgeon, now retired, Edwin Tutt Long, M.D., is presently Regional Co-Chair of the Center for Practical Health Reform and serves on the national advisory panel of the Center for Practical Health Reform. In Kansas City, Dr. Long is Curriculum Advisor and Distinguished Lecturer for the Helzberg School of Management, Rockhurst University, as well as Chairman of the Board of Roxbury Press, Inc.

A dedicated medical professional and visionary, Dr. Long served in the following positions while in private practice:

- Board of Directors, TeleMed Technologies International, Inc.
- Allegheny Cardiovascular Surgical Associates
- Senior Attending Staff, Division of Surgery, Department of Thoracovascular Surgery, Western Pennsylvania Hospital, Pittsburgh, Pennsylvania
- Founding Member and Board of Directors, West Penn Physicians Organization, Inc.
- Board of Directors, West Penn Cares, Inc.
- Attending Surgeon, Thoracic and Cardiovascular Surgery, The Pennsylvania Hospital
- Chairman, Department of Surgery, Watson Clinic
- Chairman, Department of Surgery, Lakeland General Hospital
- Attending Surgeon, Suburban Cook County Tuberculosis Sanitarium.

Inventor of four medical devices, Dr. Long is author or co-author of more than 20 contributions to professional medical journals, including *Archives of Surgery*, *American Journal of Surgery*, *Surgery Gynecology and Obstetrics*, *Journal of Thoracic Surgery*, and *Neurological Surgery*.

Dr. Long and his wife Mary live in Kansas City, Missouri, where they enjoy frequent visits from their children and grandchildren.

Ann Buzenberg, M.S.

Ann Buzenberg has edited numerous publications for business, legal and medical professionals and for the general public. Author of a children's book, *Wriggly's Rainbow*, and a parents' guide, *Learning Disabilities: Your Child and You*, she currently designs research forms for an academic medical research institute, writes on a variety of subjects, and lives in Durham, North Carolina.